T0305829

VISIONS for the FUTURE

Towards More Vibrant, Sustainable and Smart Cities

VISIONS FOR the FUTURE

Towards More Vibrant, Sustainable and Smart Cities

Editors

Thomas Menkhoff
Siew Ning Kan
Kevin Chuen Kong Cheong

Singapore Management University, Singapore

World Scientific

NEW JERSEY · LONDON · SINGAPORE · BEIJING · SHANGHAI · HONG KONG · TAIPEI · CHENNAI · TOKYO

Published by

World Scientific Publishing Co. Pte. Ltd.

5 Toh Tuck Link, Singapore 596224

USA office: 27 Warren Street, Suite 401-402, Hackensack, NJ 07601

UK office: 57 Shelton Street, Covent Garden, London WC2H 9HE

National Library Board, Singapore Cataloguing in Publication Data
Name(s): Menkhoff, Thomas, editor. | Kan, Siew Ning, 1961–   editor. |
     Cheong, Kevin Chuen Kong, editor.
Title: Visions for the future : towards more vibrant, sustainable and smart cities / editors,
     Thomas Menkhoff, Siew Ning Kan, Kevin Chuen Kong Cheong.
Description: Singapore : World Scientific Publishing Co. Pte. Ltd., [2024]
Identifier(s): ISBN 978-981-12-9309-2 (hardcover) | 978-981-12-9310-8 (ebook for institutions) |
     978-981-12-9311-5 (ebook for individuals)
Subject(s): LCSH: Smart cities--Social aspects. | Sustainable urban development. |
     Technological innovations--Social aspects.
Classification: DDC 307.760285--dc23

British Library Cataloguing-in-Publication Data
A catalogue record for this book is available from the British Library.

Cover Image: Chong Chen Kwan

For any available supplementary material, please visit
https://www.worldscientific.com/worldscibooks/10.1142/13836#t=suppl

Desk Editor: Kura Sunaina

Typeset by Stallion Press
Email: enquiries@stallionpress.com

Printed in Singapore

About the Contributors

Esther AN has been a pioneering Green Building and Sustainability Advocate for over two decades and has been instrumental in establishing CDL's leadership in sustainability. CDL has been ranked the top real estate company on the Global 100 Most Sustainable Corporations in the World since 2020 and became the first real estate conglomerate in Southeast Asia to sign the World Green Building Council's Net Zero Carbon Buildings Commitment in February 2021. Named by Reuters as one of 25 trailblazing women leaders in the fight against climate change in 2023, Esther's pioneering ESG initiatives include publishing the first sustainability report in Singapore in 2008 and issuing the first green bond by a Singaporean company in 2017. A forerunner in embracing the UN SDGs, Esther was conferred the 2018 SDG Pioneer award for Green Infrastructure and A Low Carbon Economy by the UN Global Compact initiative.

Nicole BREMSTALLER holds a Bachelor's degree in International Business Administration from Vienna University of Business and Economics with a specialisation in Business Communication, Accounting, and Taxation. Since fall 2023, she is pursuing a Master of International Management (CEMS) degree.

Kevin CHEONG has spent over 20 years in placemaking and tourism development across the Asia-Pacific. He is also an Adjunct Faculty Member at Singapore Management University, Singapore Institute of Technology, the University of Newcastle (Australia), and Coventry University (United Kingdom). He holds a Doctor of Business Administration (DBA) degree from Singapore Management University.

DENG (Frank) Xingyuan graduated from the Master of Business Management programme from Hong Kong Polytechnic University. He is currently a Public Relations Project Executive. He has great interest in market research and brand building with regional focus on the Greater China Area.

Surianarayanan GOPALAKRISHNAN is the Senior Vice President for a diversified business group headquartered in Singapore. With in-depth expertise in the international trade of industrial products and raw materials, he has successfully led and managed business verticals while handling diverse roles like category management, business development, strategy, product management, sales, and global sourcing in a professional career spanning over 25 years. Surianarayanan Gopalakrishnan holds a Doctor of Business Administration (DBA) degree from Singapore Management University, focusing on the impact of Industry 4.0 on the business models of small and medium-sized manufacturing enterprises (SMEs) in Singapore. Prior to this, he completed his Master of Business Administration (MBA) degree from Melbourne Business School (Australia) and graduated with a Bachelor of Engineering (BE) degree from the National Institute of Technology, Trichy (India).

Bernie Grayson KOH is an Assistant Director at the Centre for Teaching Excellence at Singapore Management University. He supports the University's Technology-Enhanced Learning initiatives by coordinating the design, development, and implementation of Technology-Enhanced Learning projects involving games, simulations, virtual/augmented reality, and other web-based applications.

Bernie graduated from the National Institute of Education, Singapore, with a Master of Education degree.

KAN Siew Ning started his career as an engineer with the Defence Science Organisation (DSO) in 1985 and has worked in several government agencies including IDA and DSTA. He was Director of the Police Technology Dept. from 2004–2011. As an Adjunct Faculty Member, he taught in the NTU MSc (Knowledge Management) programme for more than a decade, as well as in NUS CNM and ISS, and he currently teaches in the SMU business and IT schools. He has received three adjunct teaching awards from SMU. He has also taught in SUSS and SIT. Siew Ning was a former President of IKMS (Information and Knowledge Management Society), and has implemented C4I technology projects during his government career that spanned 27 years. He has also managed Knowledge Management (KM) projects in IDA, DSTA, and SPF. He holds BSc and MSc degrees from the National University of Singapore. He is the author of Practical Knowledge Management (2013) and co-editor of Living in Smart Cities (2018).

Kamille Storm KASTRUP is pursuing a Bachelor of Anthropology degree at Aarhus University, Denmark. She has a major interest in social negotiation between people. In her spare time, she organises annual team and leadership courses for youth.

Lily KONG is the fifth President of Singapore Management University (SMU), and the first Singaporean to lead the 21-year-old university. She was previously Provost of SMU, Vice Provost and Vice President at the National University of Singapore (in various portfolios), and Executive Vice President (Academic) of Yake-NUS College. An award-winning researcher and teacher, Professor Kong is internationally known for her research on social and cultural change in Asian cities, focusing on a range of issues such as religion, cultural policy, creative economy, urban heritage and conservation, and smart cities. In a global study by Stanford University (2020), she was identified as being among the world's top 1% of scientists in the field of Geography. Professor Kong was conferred the Singapore Public Service Star (BBM) in 2020 and Singapore Public Administration Medal (Silver) in 2006.

LEE, Jane Yeonjae is currently a User Experience Researcher at Meta and lives in San Jose, California, with her husband and two sons. She has authored three books: Moving Towards Transition: Commoning Mobility for a Low-carbon Future (2022); The 1.5 Generation Korean Diaspora: A Comparative Understanding of Identity, Culture, and Transnationalism (2021); and Transnational Return Migration of 1.5 Generation Korean New Zealanders: A Quest for Home (2018). One of her most cited articles is 'Seeking affective health care: Korean immigrants' use of homeland medical services' in Health and Place (2010). Jane has worked across numerous universities such as Singapore Management University, Stanford University, and Northeastern University, and has a PhD in Geography from the University of Auckland.

MA Kheng Min is Senior Lecturer of Organisational Behaviour and Human Resources and Director, Undergraduate Admissions, Lee Kong Chian School of Business, Singapore Management University. She obtained a PhD in Gerontology from Monash University, Australia, and holds an MA in Business Analysis from the University of Lancaster, United Kingdom. Her research interests include gerontology, ageing, learning, and the well-being of third-agers in Singapore. She is an active volunteer with many charitable organisations in promoting active ageing and the well-being of third-agers in Singapore.

Thomas MENKHOFF is Professor of Organisational Behaviour and Human Resources (Education) at SMU's Lee Kong Chian School of Business, which he joined in 2001. He is a former Board Member of the Lien Centre of Social Innovation (SMU). He also served as Academic Director of SMU's MSc in Innovation Programme. Thomas has published numerous articles in scholarly journals and several books on the socio-cultural dimensions of managing knowledge transfer, innovation, and change in multi-cultural contexts and

Asian entrepreneurship. Two of his recent publications are 'China's Belt and Road Initiative — Understanding the Dynamics of a Global Transformation,' New Jersey: World Scientific Publishing, 2020 (with Chay Yue Wah and Linda Low) and 'Catalyzing Innovations for a Sustainable Future,' New Jersey: World Scientific Publishing, 2021. In 2009, he was the recipient of SMU's university-wide 'Most Innovative Teacher Award.' In 2017, he was awarded the SMU Faculty/Staff Contribution to Student Life Award (Individual).

Ryan K. MERRIL is the Founder and Executive Secretary of the Global Mangrove Trust, a Singapore-based non-profit that works as a developer of large mangrove restoration projects in Southeast Asia. Global Mangrove Trust and its constellation of partners are using advanced digital technology to improve the accuracy of and scale up the issuance and verification of blue carbon credits. Ryan is also Co-Founder of Handprint Tech where he works as Chief Impact Officer. Before becoming a full-time entrepreneur, Ryan worked as postdoctoral research fellow in sustainability and innovation at the Lee Kong Chian School of Business at Singapore Management University. Ryan holds a PhD in Public Policy and Management with a focus on environmental policy from the Sol Price School of Public Policy at the University of Southern California and a Master's degree in Pacific and International Affairs from the University of California.

NEO Wei Leng is Senior Manager at the Centre for Teaching Excellence at Singapore Management University. She supports the Centre's efforts in promoting scholarly teaching and learning initiatives in SMU. Her research interests include learning theories, game-based learning, and educational innovations. Wei Leng graduated with a Master of Education (Learning Sciences and Technology) degree from Nanyang Technological University.

Waltraut RITTER is an independent Researcher specialising in knowledge society topics in the broadest sense. She holds an MA in Information Science and Sociology from the Free University of Berlin and an MBA from Anglia Ruskin University, Cambridge, UK. She is a Board Member of the International Council on Knowledge

Management, Member of the Euro-Asia Management Studies Association, and a Founding Member of the New Club of Paris. She holds teaching and research assignments at various universities, and has lived and worked in Stockholm, Singapore, and Hong Kong. She is now based in Berlin.

Simon SCHILLEBEECKX is an Assistant Professor of Strategy and Entrepreneurship at the Lee Kong Chian School of Business at SMU. His research focuses on the intersection of digitisation and sustainability. Beyond academia, he is the Co-Founder and Chief Strategy Officer of Handprint, a Regeneration-as-a-Service platform that helps companies grow with the planet by integrating and automating regeneration into their daily business processes. Simon is also Co-Founder of the Global Mangrove Trust. Prior to his move to Singapore in 2015, Simon obtained a PhD in Management from Imperial College London, worked in sustainable innovation consulting, and read Corporate Social Responsibility (MA, Nottingham University) and Commercial Engineering (BSc and MSc, Catholic University of Leuven).

Annika SEE is pursuing a Bachelor of Commerce degree at the University of British Colombia, specialising in Finance. Annika hopes to travel the world and see Smart Cities come to life.

Jayarani TAN is a Principal Lecturer at Singapore Management University (SMU). Besides teaching at the undergraduate level, she has conducted workshops at the post-graduate level and trained senior-level managers through SMU's executive education programme. Her research interest centres on game-based learning as she uses a gamification app as well as a simulation game that she co-developed with the SUTD GameLab to enhance experiential learning in her classroom, mostly for her undergraduate teaching.

Lydia Ying Qian TEO was a Research Assistant at Lee Kong Chian School of Business from 2021–2022. She graduated with a Bachelor of Social Sciences degree from Singapore Management University.

Her research interests include technology innovation and technology-enhanced learning.

Orlando WOODS is an Associate Professor of Geography and Lee Kong Chian Fellow at the College of Integrative Studies, Singapore Management University. His research interests include, but are not limited to, religion and multiculturalism, infrastructure development, and digital transformation in Asian cities. He holds BA (Hons) and PhD degrees in Geography from University College London and the National University of Singapore, respectively.

Caroline WONG is the Associate Dean for Learning and Teaching and Senior Lecturer, Business at James Cook University, Singapore. Her research in knowledge management takes on a multidisciplinary approach that extends into knowledge-based cities, smart cities, creative cities, and creative industries with a special focus on Singapore. In recent years, her research has also focused on applied topics like food security, luxury goods consumption, and waste management. Her academic research also extends into the scholarship of teaching and learning in higher education with a focus on work-integrated learning, student support, as well as future trends in higher education.

Irene ZHANG is pursuing a Bachelor of Commerce degree at the University of British Colombia, specialising in Marketing and Organisational Behaviour and Human Resources. With an interest in UI/UX design, she is passionate about storytelling and is always looking for opportunities to explore the intersection between business and creativity.

Contents

About the Contributors   v

*Introduction: Visions for the Future — Towards More
        Vibrant, Sustainable, and Inclusive Smart Cities*   xvii

Chapter 1   Singapore's Approach Towards Developing
             Vibrant Urban Innovation Spaces   1

             *Thomas Menkhoff, Caroline Wong, and
             Waltraut Ritter*

Chapter 2   Planning Ideas for an Educational Living Energy
             Lab — A Student Project Assignment in a Smart
             City Course   35

             *Nicole Bremstaller, Deng (Frank) Xingyuan,
             Kamille Storm Kastrup, Annika See, and Irene Zhang*

Chapter 3   Industry 4.0 Adoption by Small and Medium-Sized
             Manufacturing Enterprises in Singapore:
             A Case Study   51

             *Surianarayanan Gopalakrishnan and
             Thomas Menkhoff*

Chapter 4   AI Competency Acquisition Online? Engaging
             Undergraduate Students in an AI 101 Course
             through a Chatbot Workshop   71

             Thomas Menkhoff and Lydia Ying Qian Teo

Chapter 5   Can Gamification Help Undergraduate Students
             Acquire Leadership and Team-Building Skills?
             Findings from Two Game Projects   93

             *Jayarani Tan, Thomas Menkhoff, Neo Wei Leng,
             Bernie Grayson Koh Teck Chye, and
             Lydia Ying Qian Teo*

Chapter 6   Sensitising Students in Higher Education to
             Appreciate the Importance of Digital
             Sustainability for Smart Cities   123

             Thomas Menkhoff and Kan Siew Ning

Chapter 7   CDL — A Business Case on Building Smart
             and Sustainable Cities   151

             Esther An

Chapter 8   The Handprint Approach   171

             Simon J. D. Schillebeeckx and Ryan K. Merrill

Chapter 9   Enhancing the Sustainability DNA of
             Singapore's Gardens by the Bay Through
             Induction Training   185

             Kevin Cheong and Thomas Menkhoff

Chapter 10   Successful Ageing in 'Smart' Singapore   199

             Ma Kheng Min

Chapter 11   Towards More Inclusive Smart Cities:
             Reconciling the Divergent Logics of Data
             and Discourse at the Margins   223

             Jane Yeonjae Lee, Orlando Woods, and Lily Kong

Chapter 12 Smart Cities: A Review of Managerial Challenges
and a Framework for Future Research   245

Thomas Menkhoff

Chapter 13 Nurturing Youth Climate Action: A Blue
Carbon Perspective   295

Thomas Menkhoff and Kevin Cheong

Introduction: Visions for the Future — Towards More Vibrant, Sustainable, and Inclusive Smart Cities

Thomas Menkhoff, Kan Siew Ning, and Kevin Cheong

1. Background

As the world continues to urbanise at an unprecedented rate, cities are grappling with the challenges of managing growing populations while ensuring sustainable economic growth, social equity, and environmental protection. In response to these challenges, many cities have adopted the 'Smart City' concept, leveraging the latest technologies to enhance urban living and sustainability.

Despite the popularity of the smart city discourse, there is no universally accepted definition of a 'Smart City.' Technology firms, consultants, city governments, academics, and activists all use different frameworks and catchwords — albeit with a common core comprising the importance of good urban governance, innovative technology as an enabler for improving the quality of life, civic

> **ITU's Definition of a Smart City:** "A smart sustainable city is an innovative city that uses information and communication technologies (ICTs) and other means to improve quality of life, efficiency of urban operation and services, and competitiveness, while ensuring that it meets the needs of present and future generations with respect to economic, social, environmental as well as cultural aspects" (Recommendation International Telecommunication Union/ITU-T Y.4900).
> *Source:* https://www.itu.int/en/ITU-T/ssc/united/Pages/default.aspx.

engagement, and sustainable economic development. By focusing on liveability, sustainability, and inclusiveness, a smart city can promote the well-being of its citizens, reduce its environmental impact, and create a more equitable and inclusive community.

According to the Sustainable Cities Initiative of the World Bank, sustainable cities are "resilient cities that are able to adapt to, mitigate, and promote economic, social, and environmental change." The regeneration of places of historic significance, energy efficiency, and the reduction of the negative environmental impacts of waste generation are examples of the World Bank's sustainable development themes.

The term sustainable development refers to "development that meets the needs of the present without compromising the ability of future generations to meet their own needs" according to the 1987 UN Brundtland Report 'Our Common Future.' The report recognised the important role of Gro Harlem Brundtland (the 29th Prime Minister of Norway) as Chair of the World Commission on Environment and Development.

> **Sustainable Development Goal 11** (UN, Department of Economic and Social Affairs, Sustainable Development): Make cities and human settlements inclusive, safe, resilient, and sustainable.

According to Arcadis, a globally operating design, engineering, and management consulting firm, urban sustainability can unlock prosperity in cities, creating a sense of socio-economic security for the immediate and foreseeable future as a result of a resilient urban infrastructure, a good quality of life, social equity, inclusive opportunities for people, and environmental sustainability. The Arcadis Sustainable Cities Index ranks 100 global cities on three pillars of sustainability: Planet (environmental), People (social), and Profit (economic). In terms of the best environmental scores (Planet Pillar), European and Scandinavian cities such as Oslo and Paris stood out in 2022, while the People Pillar with its focus on health, safety, and income equality was topped by European and Asian cities such as Glasgow and Seoul (see Table 1).

Denmark's capital Copenhagen is known for its proactive approach to climate change adaptation. The city has established a Climate Adaptation Knowledge Center, which serves as a hub for collecting and disseminating climate data, research findings, and best practices. The Center collaborates with universities, research institutions, and international organisations to analyse climate projections and identify potential risks. Through this knowledge-sharing platform, Copenhagen has been able to develop effective strategies for managing sea level rise and storm surges, such as the construction of innovative flood barriers and the incorporation of green infrastructure along the waterfront. Cities which ignore planetary boundaries such as climate change or biodiversity loss are doomed to decline, resulting in rapid decay and perhaps urbicide.

Table 1. The Arcadis Sustainable Cities Index 2022: Top 5 Rankings

Planet Pillar Top 5 Ranking	People Pillar Top 5 Ranking	Profit Pillar Top 5 Ranking
1. Oslo	1. Glasgow	1. Seattle
2. Paris	2. Zurich	2. Atlanta
3. Stockholm	3. Copenhagen	3. Boston
4. Copenhagen	4. Seoul	4. San Francisco
5. Berlin	5. Singapore	5. Pittsburgh

Source: https://images.connect.arcadis.com/Web/Arcadis/%7Be08e5cda-768d-46a3-91ce-4efe16cbfc05%7D_The_Arcadis_Sustainable_Cities_Index_2022_Report.pdf.

Table 2. Examples of ESG Factors

Environmental	Social	Governance
Urban heat island effect	Health of stakeholders and community outreach	Risk mitigation and increasing heat resilience
Air pollution	Electric vehicles and public charging stations	Accounting practices and carbon taxation
Wasted energy	Waste management and resource circularity	Collaborative governance of the 5Rs (Refuse, Reduce, Reuse, Repurpose, Recycle)

Environmental, social, and governance (ESG) principles are important catalysts for smart cities (see Table 2). By integrating sustainability, transparency, and stakeholder engagement aspects into concrete smart city projects such as big data analytics or investments into cool roofs to combat heat islands, municipal governments can create common ground amongst interested/affected parties such as city inhabitants, urban businesses, smart city service providers, and municipal employees.

To create a sustainable future, social, economic, and environmental stewardship principles are increasingly applied to digital products, services, and data. Digital sustainability refers to the ability of a city to use technology to improve its efficiency and sustainability while reducing its environmental impact. A smart city that prioritises digital sustainability will focus on reducing energy consumption, managing waste more effectively, and promoting the use of renewable energy sources. By doing so, a smart city can reduce its carbon footprint and promote a more sustainable way of living.

Inclusiveness ensures that all members of the community have access to the benefits and opportunities that technology and innovation can provide. This includes ensuring that all citizens have access to affordable and reliable high-speed internet, promoting digital literacy and education, and using technology to improve accessibility for people with disabilities. A smart city that prioritises inclusiveness will be more equitable and fairer, promoting equal opportunities for all citizens.

Liveability focuses on creating a comfortable and enjoyable living environment for residents. This includes providing clean air and water, safe and reliable public transportation, efficient waste management systems, access to healthcare and education, and ample public spaces for recreation and socialisation. A smart city that prioritises liveability will improve the physical, mental, and social well-being of its residents, resulting in a happier and healthier population.

2. Singapore's Smart Nation Programme

Singapore, a small dynamic city-state in Southeast Asia, has been at the forefront of the smart city movement, developing and implementing innovative approaches to urban planning and digital technology to improve the lives of its citizens. Singapore's smart city development process gained momentum from 2014 onwards to leverage digital technologies to transform transport, urban living, finance, education, and health. Respective initiatives include smart(er) parking solutions (e.g., Parking.sg app) and testing of autonomous vehicles; pneumatic waste conveyance systems for smart(er) towns such as Punggol New Town; contactless e-payments for hawker centres and public transport commuters (e.g., Transit Link SimplyGo); the creation of a 'Smart Financial Centre' through fintech and innovation (e.g., the FinTech Regulatory Sandbox framework by the Monetary Authority of Singapore); the one-stop portal Parents Gateway app to strengthen the school–home partnership to support children in their education journey; and the Telehealth (e.g., TeleRehab) and HealthHub portals to enable citizens to access health information and services.

As Singapore is a city, a nation, and a state, the term 'Smart Nation' is frequently used by policymakers to describe the city-state's whole-of-government effort to digitise Singapore's policy processes and urban environment. With its *Smart Nation and Digital Government Office*, the Singapore Government places people at the centre of its (digital) smart nation initiatives: "We envision a Smart Nation that is a leading economy powered by digital innovation, and a world-class city with a Government that gives our citizens the best home possible

Table 3.   Strategic Focus Areas of Singapore's Smart Nation Programme

Focus Area	Strategic Goals (Examples)
Smart Digital Governance: National Digital Identity	Allowing individuals to prove their legal identity digitally via the SingPass Mobile app, e.g., signing documents and contracts easily and securely (removes the need for physical presence and paper-based signing).
Smart Digital Economy: E-Payments	Providing seamless, secure, and integrated e-payment platforms, options for cashless payments, and integrating e-payments into business processes from end to end (regulated by the 2020 Payment Services Act). It is planned to phase out cheques from 2025 onwards.
Smart People/ Digital Society: LifeSG	Providing people access to technology so that they can effectively connect with over 40 government services with just one app (e.g., childbirth registration; government payouts; and housing): LifeSG.

Source: For more information, please refer to https://www.smartnation.gov.sg/.

and responds to their different and changing needs" (https://www.smartnation.gov.sg/why-Smart-Nation/transforming-singapore).

Focus areas of Singapore's Smart Nation Programme include digital government, digital economy, and digital society (see Table 3):

- Digital Government: To provide integrated and seamless government services and use greater intelligence for more efficient policymaking and operations.
- Digital Economy: To digitise industries to increase efficacy and create new jobs and opportunities.
- Digital Society: To provide citizens access to technology so they can embrace it and utilise technology confidently and effectively to connect with the world around them.

2.1. *Smart City Chatbots*

In 2021, Singapore's Municipal Services Office (MSO) and the Smart Nation and Digital Government Group (SNDGG) launched a new AI-powered OneService chatbot which enables residents to report municipal issues such as cleanliness, illegal parking, or animal issues in their neighbourhood to the agency in charge of the issue through WhatsApp and Telegram. Chatbots are integral to the development and success of smart cities. They can enhance citizen engagement, improve public services, promote efficiency, enhance urban mobility, ensure safety and security, and contribute to sustainability efforts. By harnessing the power of chatbot technology, cities can create more connected, efficient, and citizen-centric environments.

Chatbots can play a vital role in promoting sustainability within smart cities, assisting in energy monitoring and optimisation and helping citizens and businesses track their energy consumption, aimed at making more sustainable choices. Chatbots can provide reminders to conserve energy, suggest eco-friendly alternatives, and promote efficient resource management. They can also contribute to smart waste management by providing information on recycling processes, collection schedules, and nearby recycling centres. Through such initiatives, chatbots support sustainability efforts, contributing to a greener and more environmentally conscious city.

How Chatbots Can Contribute to the Development and Functioning of Smart Cities

Chatbots act as virtual assistants, providing citizens with information and services in real time. Citizens can interact with chatbots through various channels, including websites, mobile applications, or even voice-based interfaces. Chatbots can answer frequently asked questions, provide updates on public services, and assist with navigating government processes. By enabling easy and accessible interactions, chatbots empower citizens to engage with their city and government more effectively.

An example is Askjamie, the Singapore government's virtual assistant chatbot which connects information-seeking citizens with dozens of government agency websites such as the Ministry of Social and Family Development or Smart Nation Singapore.

Another contribution of chatbots is the improvement of public services and efficiency within smart cities. Chatbots automate routine tasks and processes, allowing for streamlined service delivery. They can assist in areas such as utility bill payments, parking permits, or reporting issues

---

(*Continued*)

(Continued)

to municipal authorities. By automating these processes, chatbots reduce administrative burden, minimise waiting times, and enhance the overall efficiency of public services. Citizens can access services anytime, anywhere, without the need for physical visits or lengthy paperwork.

An interesting case study is the Assistential Chatbot of the Barcelona Council through which citizens can access social aid-related content. For more information, please refer to https://www.raona.com/wp-content/uploads/2021/08/RN_casestudy_AJBCN_EN.pdf.

Chatbots can enhance urban mobility within smart cities. They provide real-time transportation information, including bus schedules, traffic updates, and even suggest the best routes based on current conditions. Integration with public transportation systems allows citizens to plan their journeys more efficiently, reducing congestion and improving overall traffic flow. By providing convenient and up-to-date information, chatbots help citizens make informed decisions about their transportation needs, ultimately contributing to smoother and more sustainable urban mobility.

Singapore's MRT.SG provides commuters with information they need to travel smoothly on Singapore's MRT and LRT subway rail systems.

2.2. *Value Innovation at National Level*

Over the years, Singapore's agencies have planned and created several innovative sites and precincts for knowledge-intensive operations such as science and technology parks to create value at the national level. A fine example is *one-north*, Singapore's R&D and high-technology cluster developed by JTC Corporation. With a high degree of fine mixing and layering of 'work, live, play, and learn' by land use planners, one-north has attracted numerous science and engineering firms, info-comm media companies, start-ups, learning institutes, co-working space providers, retail shops, restaurants, residents, etc., and it is rapidly becoming a truly vibrant community. While a city's architecture and infrastructure are important to make it liveable, it is the vibrancy of streets and public spaces that makes it appealing and 'loveable' according to Singapore's Urban Redevelopment Authority (URA):

"In business precincts, placemaking and the vibrancy it brings leads to the exchange and flow of ideas and has become an essential part of the innovation economy. Developers and business owners all over the world are now focusing more on the value placemaking brings" (https://www.ura.gov.sg/Corporate/Get-Involved/Shape-A-Distinctive-City/Placemaking-and-Partnership).

Incorporating various residential and retail/leisure uses into the corporate zones helps promote a sense of place which is critical for sustaining urban life.

Turning Singapore Science Parks into a Living Lab for Co-Innovation

Singapore Science Park I (opened in 1982) and II (opened in 1993) are part of the vibrant greater *one-north* community, a research, development, and technology hub in Queenstown, Singapore. The 6-storey research building The Galen (61 Science Park Road, Singapore Science Park II) contains the *Smart Urban Co-Innovation Lab*, a platform for 'like-minded industry professionals to co-innovate, collaborate and test smart solutions across various industries.' Initiated by CapitaLand and supported by the Infocomm Media Development Authority (IMDA) and Enterprise Singapore, its vision is to be a global, platform agnostic co-innovation lab that builds and enables integrated communities of players to discover, develop, demonstrate, and deploy smart city solutions sustainably.

Features of Smart Urban Co-Innovation Lab
- Living Lab to showcase technology, solutions in-action.
- Test bed to adapt, stitch co-innovation partners.
- Co-develop sustainable support infrastructure, talent, etc.
- Cross-domain platform to scout/develop new ideas, technologies, business models.
- New market opportunities.

A co-innovation project example is **SMARTLab: ST's Launch of Paid AV Bus Service Trials*** led by the Singapore Technologies Engineering (ST Engineering) group (www.stengg.com), a government-linked entity with

(Continued)

(*Continued*)

business activities in aerospace, electronics, land systems, and marine sectors. Based on a consortium approach, a key objective is to develop on-demand, off-peak, last-mile connectivity solutions.

Consortium Partners
- System integrator: ST Engineering; HD Maps: GPS Land; Operator: SMRT; Booking/payment: Zipstar
- Smart Cities platform/user base: CapitaLand

Co-Innovation Opportunities
- Migration to 5G
- Ecosystems of AV players

*https://www.straitstimes.com/singapore/transport/pay-to-ride-on-driverless-buses-in-two-areas-until-april-30

By nurturing tie-ups between public sector and private sector organisations embedded in a co-innovation ecosystem (in short, living laboratory), planners are hoping to co-develop solutions for local issues or situations, and ultimately for businesses to scale the solutions regionally and internationally: "Hopefully with these ingredients, even more companies can succeed in using Singapore as their launchpad for the region in the space of urban solutions and sustainability" (https://opengovasia.com/exclusive-fostering-public-private-sector-partnerships-in-singapore/).

3. Singapore: Top Asian City in 2023 Smart City Index

The Smart City Index (SCI) created by Swiss business school Institute for Management Development (IMD) in collaboration with Singapore University for Technology (SUTD) defines a 'smart city' as an "urban setting that applies technology to enhance the benefits and diminish the shortcomings of urbanization for its citizens." The SCI categorises the different merits of a city and draws conclusions based on the perceptions and opinions of a city's residents with regard to health and safety (e.g., satisfaction with recycling services); mobility (e.g., traffic congestion); activities (e.g., green spaces and cultural activities); opportunities to work and study (e.g., lifelong learning, job creation); and governance (e.g., corruption of city officials).

Table 4. Examples of Smart City Technologies and Projects

About 99% of Singapore's government services "are digital from end to end" according to Smart Nation Singapore (https://www.smartnation.gov.sg/).

Busan, South Korea's 2nd-largest city, has a comprehensive information highway with a smart network of approximately 1,278 km, integrated CCTV systems, free Wi-Fi zones, etc.

Amsterdam is famous for its bicycles: 38% of trips made within the greater city area are made on bicycle vis-à-vis 60% in the inner city.

Hangzhou's City Brain (a big data platform) helps to deliver and manage public infrastructure services from parking cars to healthcare.

Tianjin's smart city service operations comprise 4 distinct AI platforms:

- "The Resident Voices segment uses voice recognition technologies to inform city managers and workers about the tone of voice and general satisfaction and happiness of each citizen.

- The Resident Care aspect deals with deep learning and correlation analysis to give residents access to the resources that they need to grow, learn, and thrive.

- Sensing the City is a monitoring system that collects data about air quality, weather systems, water levels, street lighting, and more to help give the local government better eyes and ears on the urban situation.

- Enterprise Services focuses on industrial and commercial relationships, the market landscape, and other means that can help boost the local economy and make the city an attractive place to conduct business."

Source: https://www.beesmart.city/en/strategy/top-10-smart-cities-in-asia.

Berlin's StadtFarm, a subscription-based smart urban farming project (https://www.stadtfarm.de/en/), is producing fresh fish, salads, herbs, and vegetables locally and sustainably in a unique closed-cycle system (with AquaTerraPonik©).

Singapore's National Parks Board (in short: NParks), a statutory board under the Ministry of National Development, has created a self-contained software package called 'NParks Explore A Route (NEAR)' Mobile App. This augmented reality (AR)-enabled mobile application complements the Coast-to-Coast (C2C) Trail, the Round Island Route (RIR), as well as the Park Connector Network (PCN) "for an interactive curated walking experience island-wide."

Source: https://pcn.nparks.gov.sg/the-pcn-experience/near/

In 2024, the Smart City Index (which ranks 141 cities by how they use technology to address the challenges they face to achieve a higher quality of life) ranked Singapore as Asia's leading smart city and the 5th smartest in the world. Singapore's land use planning and conservation authority URA (Urban Redevelopment Authority) plays a key role in making Singapore 'a great city to live, work, and play' through long-term planning and innovation, in partnership with the community.

Urban planning (urban, merged urban regions, regional, city, and town planning) is a technical and political process concerned with the use of land and design of the urban environment, including air, water, and the infrastructure passing into and out of urban areas such as transportation and distribution networks.

Urban planning guides and ensures the orderly development of set-tlements and satellite communities which commute into and out of urban areas or share resources with it. It concerns itself with research and analysis, strategic thinking, architecture, urban design, public con-sultation, policy recommendations, implementation, and management (*Source*: http://en.wikipedia.org/wiki/Urban_planning).

At the URA's new *Smart Nation CityScape* exhibition at the URA Centre (45 Maxwell Road, Singapore 069118), visitors can learn more about Singapore's Smart Nation approach by interacting with various exhibits and displays, e.g., to appreciate digital smart city services rolled out for different age groups, including seniors who can explore interesting (gamified) green walking routes created by Singapore's National Parks Board.

3.1. *Smart City-Related Pedagogical Approaches for Engaging Students*

The following (selected) smart city-related pedagogical approaches may help students get engaged with and gain a deeper understand-ing of smart(er) cities and their various elements:

City Simulation Game: Divide the class into groups and assign each group a specific task related to urban planning. Provide them with a map of a fictional city and a set of challenges such as housing development, transportation planning, and green spaces. Each group must work together to create a comprehensive plan for their assigned task, considering factors like sustainability, community needs, and budget constraints. The use of city-building games such as SimCity: Buildit can help students construct their own utopias. After the simulation, have each group present their plan and discuss the decisions they made.

Neighbourhood Walk and Analysis: Take students on a field trip to a nearby neighbourhood or urban area. Ask them to observe and analyse various urban planning aspects, such as street layouts, public spaces, architectural styles, and infrastructure. Provide them with worksheets or journals to record their observations. After the walk, come back to the classroom for a group discussion, where students can share their findings and reflect on how the design of the neighbourhood impacts its residents and their quality of life. What issues were observed? Could smart city technologies help to make the neighbourhood smarter? If yes, in what ways? If not, why not?

Zoning and Land Use Exercise: Introduce students to the concept of zoning and land use regulations in urban planning. Divide them into smaller groups and give each group a different scenario, such as a new commercial development wanting to move into a residential area or a proposal to convert an industrial/central business district area into mixed use. Ask the groups to debate the pros and cons of each scenario and come up with their zoning and land use recommendations based on the principles of urban planning covered in class.

Community Visioning Workshop: Split the class into teams and assign each team a specific community or neighbourhood. Instruct them to research the history, demographics, and current challenges of their assigned community such as gentrification, flooding, or low recycling rate. Then, have each team conduct a visioning workshop where they brainstorm ideas to improve the community's urban design, transportation, affordable housing, social amenities, resilience, carbon footprint, etc. Encourage creativity and collaboration during the workshop, followed by each team presenting their vision to the whole class.

Transportation Planning Exercise: Provide the class with a transportation-related problem, such as traffic congestion in a particular area, the need for improved public transportation, or the lack of bicycle lanes. Divide the students into groups and ask them to come up with a comprehensive transportation plan to address the issue. They should consider various options like implementing exclusive bike lanes, improving public transit routes, or introducing carpooling incentives. Each group should present their plan, and the instructor can encourage a friendly competition to see which plan is the most effective and practical.

Provided students are allocated sufficient time for group discussions and presentations, these hands-on exercises will enhance their understanding of urban planning/smart city concepts and allow them to apply their knowledge in real-world scenarios.

4. The 'Smart' City: A Digital Utopia?

The term 'Smart City' is often defined in relation to how well a city uses digital technology to improve the efficiency of urban services and to create sustainable, liveable communities where people are co-creators of smart city policies. Building a datafied city is a contested endeavour as evidenced by Toronto's scrapped 'smart' Quayside project which raised privacy concerns among locals (Josh O'Kane 2022: Sideways: The City Google Couldn't Buy, Random House Canada). Its successful management can be very complex due to numerous challenges such as implementation problems that may arise at the governance and administration level when roles, responsibilities, and success criteria are poorly defined; in terms of budgeting, competing demands, and funding constraints; and/or at the level of collaborative stakeholder management to get buy-in from citizens, private sector organisations, and city officials so that smart city projects do not end up as impractical 'white elephants.'

Besides smart city competency requirements which might be in short supply in some city councils, there is a risk that smart city technologies such as city surveillance solutions rolled out to enable faster data transmission may also increase the reputational risk of the respective municipality deploying smart CCTV cameras due to unanticipated complexities ranging from public distrust to unexpected maintenance costs. In many US cities, facial recognition bans are now being reversed to combat crime despite intense opposition from groups.

Other cities are experimenting with smart(er), AI-powered lighting systems to reduce energy consumption and light pollution by adapting streetlights according to the weather or season. An example is Bad Hersfeld (a 30,000-population spa town) in Germany where residents were enabled to control street lighting via a smartphone app by means of a pilot 'Urban Intelligence as a Service' project conducted in collaboration with Schréder, Microsoft, Deutsche Bank, [ui!] Urban Lighting Innovations, and the Berlin Institute of Technology. Besides improving the quality of life for citizens, such innovative projects help to preserve the environment, reduce energy costs, and contribute to the fight against climate change because cities are major sources of CO_2 emissions.

5. Chapter Structure

The chapters of this book examine some of the key characteristics and human dimensions of smart cities often with reference to Singapore (and its ongoing transformation towards a Smart Nation), such as the economy, environment, people, and inclusiveness, all aimed at enhancing sustainability and liveability. How does Singapore develop 'vibrant' urban innovation spaces? How can living labs create sustainable impact? How can local small and medium-sized manufacturing firms benefit from the adoption of Industry 4.0 and what are the key challenges ahead? How can one make a business case for ESG stewardship and steer Singapore's built environment towards a sustainable future? What can institutions of higher learning do to make undergraduate students AI-ready? How can one deal with Singapore's ageing population and enable smart(er) ageing? How can one connect companies and help them combat climate change by regenerating our planet? What is the future of inclusive smart cities? These questions and more are addressed in this book.

Chapter 1, authored by Thomas Menkhoff, Caroline Wong, and Waltraut Ritter, discusses Singapore's approach towards developing vibrant urban innovation spaces powered by urban planning frameworks and inspired by territorial innovation models. Strategy goals and expected planning outcomes include 'creating economic vibrancy,' 'spatial innovation,' 'job creation,' 'heritage preservation,' 'civic integration,' 'open creative milieus,' and 'ecological renewal.' This is a tall order considering Singapore's increasing population density or the multitude of stakeholders who need to be engaged. With regard to four such spaces, namely, Singapore's Punggol Digital District, Jurong Innovation District (I4.0), One-North, and the historic Bras Basah.Bugis District, the authors shed light on some of the aspirations and challenges planners are facing to create 'a great city to live, work, and play,' a tagline which is the mission of Singapore's national land use planning and conservation agency URA.

Chapter 2, authored by Nicole Bremstaller, Deng Xingyuan, Kamille Storm Kastrup, Annika See, and Irene Zhang, presents basic planning ideas for an educational living energy lab (the chapter originated from a student project assignment in a smart city course taught by the editors). Living Labs are places where citizens, technologists, private and public sector organisations, and other stakeholders come together to co-create ideas, projects, and tools to address local challenges. The chapter discusses how the project was designed to enhance students' understanding of the importance of energy conservation, sustainability, and solar energy in a smart city context. It also highlights some of the opportunities and challenges associated with developing educational living lab projects.

Chapter 3, authored by Gopalakrishnan Surianarayanan and Thomas Menkhoff, explores how Singapore's manufacturing small and medium-sized enterprises (SMEs) are embracing Industry 4.0 based on the story of Racer Technology and its Chief Executive Officer, Mr Willy Koh. The chapter examines the factors that contribute to the adoption success of Racer Technology, including the importance of entrepreneurial leadership as well as the availability of funding, support services, and technological expertise.

Chapter 4, authored by Thomas Menkhoff and Lydia Ying Qian Teo, examines the use of chatbot workshops to engage undergraduate students in an AI 101 course. The authors developed and corroborated a conceptual model indicating how the hands-on chatbot workshop might lead to outcomes such as motivation and engagement, eventually resulting in the achievement of desired learning outcomes such as natural language processing (NLP) know-how and know-why.

Chapter 5, penned by Jayarani Tan, Thomas Menkhoff, Wei Leng Neo, Bernie Grayson Koh, and Lydia Ying Qian Teo, presents findings from two game projects that use gamification to help undergraduate students acquire leadership and team-building (LTB) skills. The chapter explains how these games can be used to enhance students' understanding of the importance of leadership and teamwork. The study contributes to the still nascent body of knowledge pertaining to effective gaming mechanics in educational contexts by sharing and comparing the design principles of 'GameLead' and 'Stranded' aimed at engaging students and achieving LTB-related learning outcomes.

Chapter 6, authored by Thomas Menkhoff and Kan Siew Ning, explores how the authors sensitised undergraduate students to appreciate the importance of digital sustainability in an undergraduate course on 'Sustainable Digital City.' The authors describe the course context and pedagogical challenges, present key teaching and learning contents and delivery approaches (aimed at enabling students to appreciate the importance of tackling issues pertaining to digital sustainability), discuss related problems, and examine the conclusions with a view towards streamlining key learning outcomes in relation to digital sustainability challenges. Student feedback suggests that 'make-it-real' projects where learners can experience the urgency of climate change in combination with carefully curated site visits can promote their sensitivity and concrete engagement with the critical issue of 'digital pollution' aimed at nudging them to realise the urgency of achieving net-zero carbon dioxide emissions.

Chapter 7, authored by CDL's Chief Sustainability Officer, Esther An, presents a business case on building smart and sustainable cities, using the best-practice example of City Developments Limited (CDL) with its ethos 'Conserving as we Construct.' CDL is a Singapore-listed globally operating real estate company with a presence spanning 90 locations in 26 countries. CDL is well known for its sustainability (ESG, integrated reporting, etc.) initiatives in its buildings and communities. Based on the case of CDL, the author explains why and how companies need to integrate ESG best practices into their business to become more efficient in reducing their carbon footprint, while also improving the lives of residents.

Chapter 8, authored by Simon Schillebeeckx and Ryan K. Merrill, features a new approach to sustainability called the Handprint approach. The authors assert that instead of focusing solely on reducing negative impacts, we should also focus on creating positive impacts, or 'handprints.' They outline a framework for measuring and maximising positive impacts in areas such as social well-being, human rights, and biodiversity. From reforestation to safe water, from ocean health to gender equality, Handprint Tech quantifies impact in real time according to United Nation's SDG.

Chapter 9, authored by Kevin Cheong and Thomas Menkhoff, focuses on the sustainability initiatives at Singapore's iconic Gardens by the Bay. The authors describe why and how the Gardens by the Bay management team initiated the implementation of an induction training programme to instil sustainability principles in their staff and volunteers. A key proposition is that induction training is critical to create a culture of sustainability in horticultural destinations, which can be applied to other organisations as well. The chapter is a reprint of a Teaching Case by Kevin Cheong and Thomas Menkhoff published by the Singapore Management University in 2022 (SMU-21-0048).

Chapter 10, authored by Ma Kheng Min, explores the concept of successful ageing in Singapore. She argues that as Singapore's population ages, it is crucial to ensure that seniors can live fulfilling and meaningful lives. The author highlights a range of initiatives that Singapore has implemented to support senior citizens, including health and wellness programmes, lifelong learning opportunities, and social support networks.

Chapter 11, authored by Jane Yeonjae Lee, Orlando Woods, and Lily Kong, examines the concept of inclusiveness in smart cities. The authors argue that while data and technology can be powerful tools for improving urban planning, management, and bottom-up smart cities, they can also create new forms of exclusion and inequality ('knowledge politics'). The authors discuss limitations of the current discourses around inclusive smart cities and suggest a need for a more nuanced definition of 'inclusiveness.' The chapter is a reprint of an article published in Wiley's Geography Compass in 2020.

Chapter 12, authored by Thomas Menkhoff, critically discusses what makes a city 'smart,' introduces the main smart city components, and examines some of the typical smart city applications as well as associated knowledge gaps. He also highlights some of the managerial challenges arising from these issues and outlines opportunities for more theory-based, empirical research on smart cities from a purpose-driven business perspective linked to human betterment. The chapter is a reprint of a contribution to the 'Handbook on the Business of Sustainability' published by Edward Elgar Publishing in 2022.

Chapter 13, authored by Thomas Menkhoff and Kevin Cheong, propagates a blue carbon perspective to make students climate resilient. The concept of blue carbon has gained recognition among scientists, policymakers, and environmentalists, but is not yet widely understood by the general public. From the perspective that more needs to be done given the urgency of climate action, the authors highlight examples of how educators and the corporate sector can partner to engage and motivate our youths into taking concrete climate action.

We hope that readers of *Visions for the Future: Towards More Vibrant, Sustainable, and Inclusive Smart Cities* will come away with a deeper understanding of the challenges and opportunities of building better cities which are liveable, sustainable, and inclusive.

Chapter 1

Singapore's Approach Towards Developing Vibrant Urban Innovation Spaces

Thomas Menkhoff, Caroline Wong, and Waltraut Ritter

1. Introduction

In 2024, the annual IMD-SUTD Smart City Index (SCI) — which ranks 141 cities by how they use technology to address the challenges they face to achieve a higher quality of life — ranked Singapore as Asia's leading smart city and the 5th smartest in the world. The SCI was conceptualised by the Smart City Observatory at the Institute of Management Development (IMD) and the Singapore University of Technology and Design (SUTD) to assess the perceptions of residents on issues related to structures and technology applications available to them in their city. Methodologically, the SCI rests on two pillars for which perceptions from residents are solicited: (1) the Structures pillar, which refers to the existing urban infrastructure, and (2) the Technology pillar, which describes the technological provisions and services available to the inhabitants. Each pillar is evaluated over five key areas: *health and safety, mobility, activities, opportunities*, and *governance*.

1

While Singapore's leaders remain steadfast in further transforming Singapore into a liveable 'city in a garden,' the city-state's innovativeness is a matter of concern. In the 2021 Global Innovation Index (GII) published by the World Intellectual Property Organization, which ranks countries based on their success and capacity in innovation, Singapore was ranked 8th (in the 2023 GII, Singapore was ranked 7th). The country took the top spot (1st) in the Innovation Input Sub-Index, and it also ranked first in two new indicators relating to capital raising and financing: 'Venture Capital Investors' and 'Venture Capital Recipients' (Intellectual Property Office of Singapore, IPOS, n.d.). The availability of risk capital, a pro-business government, talent, openness, and political stability are important ingredients for the creation of innovation clusters as evidenced by the rise of Silicon Valley, Boston's Route 128, or the Beijing Zhongguancun (ZGC) Science Park. But the success of top-down policies to create novel knowledge and technology hubs cannot be taken for granted as the failure of Bavaria's 1999 cluster initiative or Malaysia's BioValley launched in 2005 indicates.

Singapore's state-led planning approach towards creating a 'vibrant community that will work, live, play and learn together in a dynamic estate where people can congregate, collaborate and innovate' (JTC Corp, 2018) makes for an interesting case study. The Jurong Town Corporation (JTC) is a statutory board under the Ministry of Trade and Industry (MTI) that develops industrial infrastructure that supports the growth of new industries and builds vibrant, innovative developments where they can grow and thrive (JTC, 2023). As we shall outline in the following, Singapore's Punggol Digital District, Jurong Innovation District (I4.0), One North, and the historic Bras Basah.Bugis precinct are examples of such geographical spaces where 'people' (students, employees of small and big companies), institutes of higher learning (IHLs), government agencies, etc., can 'live and work together, and where they can theoretically benefit from their proximity (Boschma, 2005) to collaborate and elaborate common ideas and outputs' (IGI Global, n.d.). A closer look at such spaces helps to understand both the importance and challenges of developing 'clusters,' 'industrial districts' (Harrison, 1992), 'regional innovation systems,' and 'entrepreneurial ecosystems' as forms of territorial agglomeration of knowledge-centric organisations. Interesting discussion questions include the following: Which urban growth model underpins Singapore's urban development approach? What are some of Singapore's future growth areas? What does it take to foster collaboration and innovation between businesses, professionals, and nearby institutions due to spatial synergy effects and collaborative working practices? How can one 'create a vibrant community whose members will work, live, play and learn together in a dynamic estate where people can congregate, collaborate and innovate' (JTC Corp, 2018)?

To understand the real dynamic of a place requires high-quality mixed-methods research. The social and intangible dimensions of a city can hardly be reflected in city rankings such as the abovementioned IMD SUTD Smart City Index or the Cities in Motion Index developed by academics of IESE Business School. Despite methodological challenges with regard to objective city comparison and continuous efforts by city governments to strengthen the city's reputation, city (place) branding underscores the point that cities require concerted efforts and know-how to make them liveable. The ability to attract talents and businesses determines the competition among different places, which causes city governments as well as local and regional economic development agencies to develop such marketing plans in the first place. As Howkins (2009) has stressed, a key question for creative people and knowledge workers today is this: Where do people go when they want to think? Creating creative urban ecologies can drive a creative economy which in turn can power a 'smart(er) city.' How can Singapore manage to do that?

2. Singapore's Approach Towards Urban Development: Theoretical Perspectives

There are several disciplines which can help planners and policymakers design spatial strategies aimed at creating new commercial, industrial, and living spaces: urban planning, geography, regional sciences, operations research, urban economics, and urban sociology. Examples of urban growth models include the concentric circle growth (CCG) model, the sector growth model, and the multiple-nuclei growth

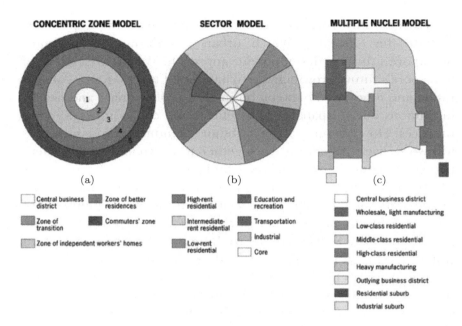

Figure 1.   Examples of Urban Growth Models

model that have been explored extensively by Chaiechi *et al.* (2020) in their analysis of the urban design and economic growth of Cairns and Singapore (see Figure 1).

The concentric circle (or zone) growth model (CCG) was developed by US sociologist Ernest Burgess in the 1920s (Burgess, 1924) to explain the distribution of social groups within the urban area of Chicago. The inner zone refers to the central business district (CBD) which comprises the principal commercial streets and economic activities as well as the main public buildings. The zone of transition contains low-density, mixed commercial (e.g., factories) and residential land. The zone of independent workers' homes is made up of high-density, lower-class housing (inner suburbs) while the zone of better residences refers to middle-class, low-density residential areas in the suburbs. The commuters' zone refers to the high-class homes of wealthy residents located in the very low-density outer suburbs who can afford to drive to the CBD for work, shopping, and leisure.

The CCG model helped to develop the so-called bid rent theory, a geographical economic theory put forward by Harvard economist and regional planner William Alonso (1968) to account for intra-urban variations in land use and rent gradients (yields) based on the notion that the more accessible an area (i.e., the greater the concentration of customers), the more profitable it is.

The CCG model influenced the development of the sector growth model by British land economist Homer Hoyt (1939) who argued that economic growth follows an outward-extending transport network along railway lines, highways, or rivers.

The multiple-nuclei growth model (MNG) was developed by urban geographers Chancy Herris and Edward Ullman (1945) who argued that cities do not grow around a single nucleus such as a CBD, but rather via a multiple-nuclei structure. Each nucleus can give rise to the creation of a smaller growth centre with certain regional areas or clusters specialising in certain business activities. In contrast to the CCG model and the sector growth model, MNG considers the importance of car ownership and automobile transportation modes (new disruptive technology in the early 1940s) as well as their land use and inter-urban connectivity implications.

2.1. *Singapore's Urban Growth Model and Growth Areas*

According to Chaiechi *et al.* (2020), Singapore exemplifies the multiple-nuclei growth model (MNG). The city-state's multiple nuclei form centres of economic activity, housing, and lifestyle attractions in five regions or urban planning subdivisions, namely, Central Region, North Region, North-East Region, East Region, and West Region (see Figure 2).

Singapore's 'Central Area' (within the Central Region) comprises the Central Business District (CBD), the Downtown Core (City Hall, Bugis, Marina Centre, Nicoll zones, Marina East, and Marina South), the Museum Planning Area, Newton, Orchard, Outram, River Valley, Rochor, the Singapore River, and Straits View (https://www.ura.gov.sg/maps/). Besides prime office space, there are several museums, theatres and art galleries, riverfront developments, hotels, retail, and (high-end) residences in this 'city area'

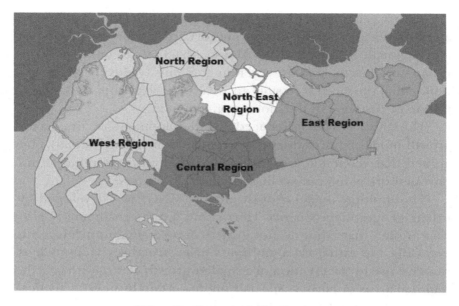

Figure 2.   Singapore's Five Regions
Source: https://en.wikipedia.org/wiki/Regions_of_Singapore.

such as Marina Bay Residences, Marina One Residences, South Beach Residences, and Wallich Residence.

One North, a subzone of Queenstown, is located in Singapore's 'Central Region.' It was first developed by JTC Corporation as the country's research and development and high-technology cluster. It aspires to become Singapore's 'Silicon Valley' (see Section 4.3).

Health City Novena is part of the Newton planning area within the central area. This cluster is an integrated community of healthcare, medical education, and translational research in a vibrant and sustainable communal environment.

The *Bidadari Estate* is an upcoming housing estate located within the Central Region of Singapore. It is part of the planning area of Toa Payoh and was once the site of the Bidadari Cemetery.

The 'North Region' comprises the Central Water Catchment, Lim Chu Kang, Mandai, Sembawang, Simpang, Sungei Kadut, Woodlands, and Yishun.

The 'North-East Region' consists of Ang Mo Kio, Hougang, North-Eastern Islands, Punggol, Seletar, Sengkang (Sengkang New

Town, Rivervale, Compassvale, Buangkok, Anchorvale, Fernvale, and Jalan Kayu), and Serangoon.

Tampines Town (located in the geographical region of Tanah Merah along the north-eastern coast of the East Region of Singapore) was the first regional centre to be developed based on the precinct concept in 1979 for the promotion of neighbourliness. In 1992, it won a UN World Habitat Award for its progressive town design. Besides public and private residences, the centre is supported by international schools, big warehouse retail parks for global brands such as Ikea, as well as the Mass Rapid Transit (MRT) system connected with bus, taxi, and private vehicle services. The name Tampines came from the Malay word for the ironwood tree found abundantly in the town, 'Tempinis.' A social innovation in Tampines Town is 'Our Tampines Hub' (OTH), Singapore's largest integrated community and lifestyle hub, bringing together multiple agencies and offering a comprehensive range of services and facilities. Led by the People's Association, OTH represents a new model of community-focused development, celebrating the full sense of community through meaningful engagement, enriching residents through multiple experiences, and empowering them to stake ownership of the hub, so as to build a robust identity for Tampines and a truly sustainable community (*Source*: https://www. pa.gov.sg/our-network/our-tampines-hub).

The 'East Region' consists of Bedok, Changi, Changi Bay, Paya Lebar, Pasir Ris, and Tampines.

Several of Singapore's heavy manufacturing areas are located in the 'West Region' comprising Boon Lay, Jurong West, and Jurong East. Jurong was chosen as the site for Singapore's industrialisation and urbanisation programmes from the 1960s to the 1980s. Future growth areas in the western area include Jurong Lake District, earmarked to become a second CBD, and Tuas Port, featuring digital innovations such as Digitalport@SG and the Just-in-Time System to streamline vessel clearance processes, enable just-in-time operations, and improve the turnaround time of ships in the port.

The Urban Redevelopment Authority (URA) considers this part of Singapore as the 'Western Gateway' that will leverage on improved transport linkages and the global maritime connectivity of Tuas Port

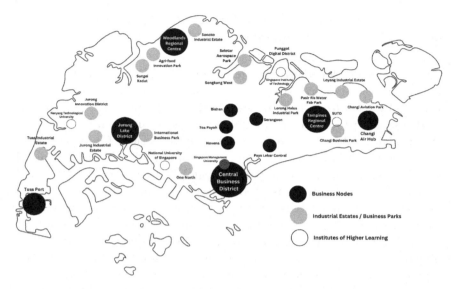

Figure 3.   Singapore's Economic Hubs and Growth Areas

Source: Adapted from https://www.ura.gov.sg/Corporate/Planning/Master-Plan/Themes/ Local-Hubs-Global-Gateways/Strengthening-Economic-Gateways — Map by Lukas Ming Menkhoff.

to facilitate the seamless movements of goods, services, and people (URA, n.d.a): "The presence of world-class universities also makes it the ideal location for businesses, academia and talent to meet, exchange ideas and collaborate" (URA, n.d.a).

The COVID-19 crisis accelerated the planning for more mixed use in the city centre facilitated by the Strategic Development Incentive (SDI) scheme to convert older office buildings into mixed-use developments in the form of hotels, residences, gyms, grocery stores, and eateries in line with the live, work, and play vision of the URA: "The area will continue to grow and become even more vibrant, by accommodating a wider diversity of uses and more jobs and business opportunities for the future economy. More city living options nearer to amenities have been planned too, with delightful streets and public spaces that celebrate our rich cultural, heritage and green assets" (URA, n.d.a).

In July 2021, the URA started public engagement for the Long-Term Plan Review (LTPR) exercise, which is done every 10 years to review the government's plans for long-term use over the next 50 years. The CBD continues to be transformed into a 'multi-faceted district,' possibly with more homes and less commercial spaces, and the reverse in some suburbs. To what extent newly released white sites with their big price tags will make it harder to turn the urban commons concept into practice so that citizens from all backgrounds and social statuses can use the city as a platform remains to be seen.

3. Singapore's Hierarchical Planning Process

Singapore is a very small country with a land area of 721.5 sq. km. Therefore, land is very precious and needs to be managed well. One lever is the city-state's systematic, long-term, and hierarchical planning approach. Table 1 illustrates the various components of the planning process from the strategic land use and transportation plan called 'Concept Plan' to the 'Development Control Plan' (https://www.ura.gov.sg/Corporate/Planning/Master-Plan).

The URA's Master Plan differentiates different types of land sites or zones. Zoning is a land use planning tool used by local

Table 1.   Singapore's Planning Process

Process	Objective
Concept Plan	The Concept Plan is the strategic land use and transportation plan that guides Singapore's development over the next 40–50 years. It is reviewed every 10 years.
Master Plan	The Master Plan is the statutory land use plan which guides Singapore's development in the medium term over the next 10 to 15 years. It is reviewed every five years and translates the broad long-term strategies of the Long-Term Plan into detailed plans to guide the development of land and property. The Master Plan shows the permissible land use and density for developments in Singapore.

(*Continued*)

Table 1. (*Continued*)

Process	Objective
Land Use Plan	The Land Use Plan outlines the land use strategies in support of a high-quality living environment and the needs of future generations.
Government Land Sale	Land is released for development to bring the Concept and Master plans to reality.
Development Control Plan	The Master Plan is supported by Special and Detailed Control Plans (SDCPs). One purpose of these development control plans is to evaluate and grant approval for development projects in line with government planning strategies and guidelines.

governments and planning agencies to designate mapped zones which regulate the use, form, design, and compatibility of development. In Singapore, there are about 30 different zones/sites such as residential, commercial, hotel, business park, white sites, health/medical, education, place of worship, and civic and community institution (https://propertyreviewsg.com/ura-masterplan-zoning-interpretation/).

Zoning may be use based (regulating the uses to which land may be put, also called functional zoning), or it may regulate building height, lot coverage (density), and similar characteristics, or some combination of these. Zoning describes the control by authority of the use of land and of the buildings thereon. Areas of land are divided by appropriate authorities into zones within which various uses are permitted. For example, in Singapore, one seldom finds an industrial park located next to an area of landed properties because zoning discourages it.

Table 2 shows the various agencies and their roles in terms of planning and managing land and buildings in Singapore.

4. Four Collaborative Urban Innovation Spaces

In the following, we describe four examples of future-oriented collaboration spaces for innovation and discuss some of their planning imperatives and expected outcomes.

Table 2.   Agencies involved in Planning and Managing of Land and Buildings

Agency	Function
Housing and Development Board (HDB)	Singapore's public housing authority. HDB plans and develops Singapore's housing estates, building homes and transforming towns to create a quality living environment for all. The HDB also provides various commercial, recreational, and social amenities 'in our towns for our residents' convenience.'
Urban Redevelopment Authority (URA)	Land use planning and conservation authority. The URA's mission is to make Singapore a great city to live, work, and play: "We strive to create a vibrant & sustainable city of distinction by planning and facilitating SG's physical development."
Singapore Land Authority (SLA)	A statutory board under the Ministry of Law. The SLA's mission is to optimise land resources for the economic and social development of Singapore.
JTC Corporation (JTC)	Lead agency to spearhead planning, promotion, and development of Singapore's dynamic industrial landscape. Since 1968, JTC has played a major role in Singapore's economic development journey by developing land and space to support transformation of industries.
Building and Construction Authority (BCA)	Champions the development of an excellent built environment for Singapore. The term 'built environment' refers to buildings, structures, and infrastructure in our surroundings.

4.1. *Punggol Digital District (PDD): The Next-Generation Smart and Integrated District*

Punggol is a systematically planned new town area located on the Tanjong Punggol peninsula in the North-East Region of Singapore (URA, n.d.). The transformation of this old settlement into a new model of housing began in 1998 under the Punggol 21 initiative. Its development into a modern waterfront town is based on the Punggol 21-plus plan.

In 2018, a new master plan for a 'smart and sustainable mixed-use' 50-ha **Punggol Digital District** was announced which was to progressively open from 2023 onwards according to Singapore media reports (*Straits Times*, 2020; 2021a). The master plan was developed by JTC Corporation in alignment with Singapore's Smart Nation and digital economy plans. Confirmed tenants include the Singapore Institute of Technology, a new technological institution that will enrol more than 10,000 students to be taught by 500 academic staff, smart living solutions firm Delta Electronics Int'l (Singapore), and robotics design company Boston Dynamics.

In support of Singapore's transition towards a Smart Nation, 'PDD will be the showcase of how integrated master planning and technology can help create a more liveable and sustainable environment for the community at a local district level, and foster a thriving business and lifestyle ecosystem that attracts talent and enables innovation.' The following are examples of district-wide infrastructure and services that will be available in the area:

- "An Open Digital Platform that will integrate the management of the various buildings within a single estate system as well as centralise the data collection of these systems. Innovators can plug into the platform and access the data to create community-centric solutions.
- A Centralised Logistics Hub that will serve as a single stop where all goods can be dropped off and picked up, improving productivity and reducing traffic.
- A District Cooling System to centralise cooling needs with space and cost savings that will reduce the district's carbon footprint.
- Pneumatic Waste Collection (with a district-wide underground vacuum-pipe network) to eliminate the need for waste collection trucks and eradicate odour from refuse chutes.
- A Smart Energy Grid to enable consumers to adopt clean sources of energy for daily use (e.g., charging electric vehicles) and to facilitate greater energy efficiency plus savings (e.g., through smart metering)."

Source: https://www.ura.gov.sg/Corporate/Planning/Master-Plan/Urban-Transformations/Punggol-Digital-District.

The Government expects that about 28,000 digital economy jobs to be created within the district in future growth areas such as artificial intelligence and cyber security: "The entire Punggol Town with its housing precincts and the Digital District will serve as a living lab, for public agencies and companies to test out new ways of living, working and delivering services," said Mr Teo, Singapore's Coordinating Minister for National Security, during the ground-breaking ceremony for the district (*Straits Times*, 2020).

The centre piece of the digital district will be a software platform called the Open Digital Platform (ODP) developed by JTC, the Government Technology Agency GovTech (a statutory board under the Prime Minister's Office), and ST Engineering to not only optimise building management and resources but also ensure the seamless communication of smart system components such as autonomous vehicles, lifts, or district cooling systems powered by Open Standard Multiprotocol Middleware.

Punggol Digital District will have a digital twin based on the real-life 3D model of PDD enabling technology firms interested in test-bedding 'their solutions to plug directly into the digital twin and experiment without worrying about compromising either their solutions or the district' (JTC, 2022). By collecting real-time data about the weather, power consumption, and the number of people in the district through sensors and individual systems across the estate, the ODP can predict energy demand and supply in support of effective energy maximisation: "The data from these disparate systems will be fed into the machine learning models of the ODP. For example, it will be able to adjust air-conditioning output automatically and save energy without compromising user comfort" (JTC data scientist Miss Shermaine Wong as quoted in JTC (2022)). By using AI, it is envisaged that Punggol as "a tech-enabled smart and sustainable district" can achieve 30% energy savings in comparison to standard buildings.

As an AIOT-powered living lab with real-time data on transport conditions, energy and water consumption, noise, rainfall, etc., public agencies, companies, and students can test green technologies, sustainable urban solutions, and public services. As highlighted by Mr Teo, the use of data analytics is expected to improve planning and town management: "For example, we can study footfall data to find the best places to locate community facilities and retail spaces that most conveniently serve our residents" (*Straits Times*, 2020).

Planners hope to create an internally and externally 'well connected' ecosystem/growth cluster of knowledge-intensive stakeholders such as digital and cyber-security firms with "close integration of industries, academia, and the community at large" embedded in an "attractive and pedestrian-friendly physical environment" (URA, n.d. — PDD Design Guidelines).

By locating some of SIT's research labs and learning facilities in JTC's business park buildings while housing corporate research and development facilities and start-up spaces within in the SIT campus, the intent is to foster a close collaboration between students and businesses. According to Associate Professor Steven Wong, SIT's director of projects in the provost's office, "... when we eventually move to Punggol, our students will find working with industry in real-world environments second nature" (*Straits Times*, 2020). Several of the business park buildings will be connected to the SIT campus via an elevated pedestrian walkway, like the one that links IMM mall, the Devan Nair Institute for Employment and Employability, and Ng Teng Fong General Hospital in Jurong East. To what extent these walkways will lead to the 'cross-fertilisation of ideas' and 'networked innovation' remains to be seen.

PDD aspires to become "an inclusive and green lifestyle destination for the surrounding community. It is envisioned to be:

- A community playground and green heart for all residents of Punggol
- A car-lite precinct that allows everyone to travel with ease
- A vibrant economic and learning hub
- A pilot Enterprise District, with shared spaces between industry and academia"

Source: https://www.ura.gov.sg/Corporate/Planning/Master-Plan/Urban-Transformations/Punggol-Digital-District.

4.2. *Jurong Innovation District — Creating a Beacon and Synergistic Ecosystem for Industry 4.0 Collaborations*

Spread over 600 hectares, Singapore's visionary **Jurong Innovation District** (JID) is master planned and developed by JTC (see Figure 4).

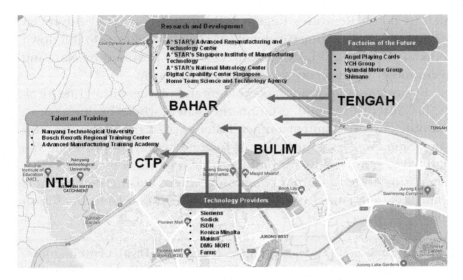

Figure 4. Jurong Innovation District

It aims to become an 'ecosystem of leading players across the entire industrial value chain' with emphasis on advanced manufacturing. By locating research institutes, training providers, technology experts, capability developers, and advanced manufacturers in 'one convenient spot,' the planners are hoping to make Industry 4.0 'collaborations that much easier.' Committed companies include Shimano's factory of the future and Hyundai Motor Group's Innovation Centre, an 'open innovation base that researches and demonstrates customer's entire mobility lifecycle value chain' (Hyundai Motor Group Website).

JTC's Industry Connect Office @ Jurong Innovation District is envisaged to be a 'one-stop centre to help more manufacturers, especially local SMEs, accelerate their Industry 4.0 transformation' (JTC, 2021c).

Tenants include McKinsey's Digital Capability Centre, the Advanced Remanufacturing and Technology Centre (ARTC), and the Singapore Institute of Manufacturing (SIMTech). The latter two are affiliated to Singapore's Agency for Science, Technology and Research (A*STAR). Industry 4.0 solution providers such as Siemens, Makino, Sodick, and Konica Minolta, as well as training providers such as Bosch Rexroth's Regional Training Centre and Advanced Manufacturing Training Academy (AMTA) have agreed to put down roots in JID. Critical talent resources will be provided by Nanyang Technological University, which is located nearby.

By situating important stakeholders and players within one locality, JTC intends to build 'a unique ecosystem where businesses can innovate and grow, and at a much faster pace.' Contrary to the past, when Singapore's statutory boards focused largely on providing the right space and conducive buildings for tenants, the current challenge is to 'foster a vibrant community where like-minded partners come together' aimed at fulfilling Singapore's Manufacturing Vision 2030.

> "Jurong Innovation District (JID) is a 600-hectare District covering the Nanyang Technological University, JTC's CleanTech Park, as well as the Bulim, Bahar and Tengah areas. The district will play host to an advanced manufacturing campus, factories of the future and companies at the forefront of key growth sectors such as advanced manufacturing, urban solutions and smart logistics."
>
> *Source*: https://infrastructure.aecom.com/2019/singapore/jurong-innovation-district.

JTC hopes that many businesses will tap into JID's advanced 4.0 manufacturing ecosystem to further grow their capabilities in collaboration with other partners. An example is Flexmech Engineering, a local business supplying precision equipment and services to clients across different sectors globally, which collaborated with Bosch Rexroth to create an Industry 4.0 'lite' solution for Flexmech's clients: "Through this, the longevity of the machines are extended, which translates to potential savings of up to $250,000 annually" (https://www.jtc.gov.sg/about-jtc/news-and-stories/feature-stories/collaborating-for-industry-40-success).

Another example is the collaboration between the Advanced Remanufacturing and Technology Centre (ARTC), Procter and Gamble (P&G), and local SME JM Vistec to develop an Industry 4.0 solution for SK-II, a well-known P&G brand: "Through this collaboration, an automated inspection system powered by artificial intelligence was created to inspect the glass bottles used in some of SK-II's products" (https://estates.jtc.gov.sg/jid/stories/jm-vistec). The collaboration project has resulted in faster inspections of the P&G packing line.

Siemens partnered with A*STAR and the National Additive Manufacturing Innovation Cluster (NAMIC) to develop 3D printed face shields for use on the COVID-19 frontlines (https://estates.jtc.gov.sg/jid/stories/shielding-the-frontlines-with-3d-printing).

A*STAR helped to develop the final product with the right medical coating; through NAMIC, Siemens connected with partners such as HP and Jabil, who provided the machines and technical expertise required for large-scale 3D printing.

A key JID player is YCH Group's Supply Chain City, an innovative 'premier supply chain nerve centre.' With more than $200 million in building investment and covering more than 6.5 hectares of land, Supply Chain City has been designed for scalability and operations around the clock, envisioned to improve logistics productivity with the help of new technologies such as advanced robotics/AI and to contribute to greater local, regional, and international connectivity.

4.2.1. *Developing a framework for connections and a mindset for collaboration*

The above-mentioned examples underline the importance of developing a 'framework for connections' within the JID ecosystem to nurture a collaborative mindset to achieve common objectives and outcomes. One practical approach is JTC's 2021 Industry Connect initiative (https://estates.jtc.gov.sg/jid/industry-connect) which resulted in the participation of more than 1,000 businesses (operating in JTC's estates) in various Industry 4.0 outreach programmes. According to official estimates, 250 SMEs are undergoing various phases of digitalisation aimed at transforming their businesses and seizing new opportunities.

To help SMEs embark on the Industry 4.0 journey, Singapore's government and industry partners have provided several schemes and grants. Transformation benefits include reduced operating costs, redesigned workflow and processes, an increase in machine utilisation rate and productivity, as well as higher sales and higher margins. To develop a strong talent pipeline, the agency has organised Industry Days such as Biopharma Industry Day, Electronics Day, and Aerospace Day for students.

Another collaboration tool set up under the Singapore Business Federation's Industry 4.0 Human Capital Initiative Enabler Programme is *Communities of Practice* such as the partnership with Siemens and Bosch Rexroth aimed at helping interested companies with the scaling and deployment of Industry 4.0 solutions: "This forms part of the continual support structure put in place to uplift businesses and reach the intended outcomes. In the meantime, we are also creating platforms and formalising partnerships to advance sustainable manufacturing in Singapore and Asia" (JTC, 2021a).

> The goal of Singapore's Manufacturing 2030 vision is to grow the manufacturing sector (currently valued at $106 billion) by 50%. The contribution of manufacturing to Singapore's GDP will be maintained at around 20%. "We hope to undertake this growth through a focused, three-pronged approach: continue investments in the manufacturing ecosystem; build on our existing research capabilities; and promote Industry 4.0 adoption. What we want to achieve is for Singapore to become a global business, innovation, and talent hub for advanced manufacturing. Our aspiration is for the country to be a place where industrialists can come together and push the boundaries of innovation. And to meet this target, collaborations — not only between the private and public sectors, but also among the vertical and horizontal spheres within the private sector — will be critical" (JTC, 2021a).

4.3. *One-north: Creating a Vibrant Community That Will 'Work, Live, Play and Learn' Together*

One-north is a 200-hectare estate and a subzone of Queenstown in Singapore — first conceptualised in 1991 as part of the National Technology Plan. In 2001, it was launched as part of Singapore's research and development and high-technology cluster and jointly developed by JTC Corporation alongside other agencies and private developers.

The one-north master plan envisions a 'vibrant' community that will 'work, live, play and learn' together in a dynamic estate where people can 'congregate, collaborate and innovate' (JTC Corp, 2018). Urban planning strategies such as clustering of companies and businesses with similar activities serve to actualise the vision as proximity foster collaboration and innovation between businesses and professionals, and companies are also able to leverage and benefit from synergies with nearby business parks and institutions.

Branded as the local, regional, and global centre for high technology and high-tech innovation, one-north is strategically located

Figure 5. Fusionopolis (Photo credit: Thomas Menkhoff)

close to renowned education and research institutions such
as National University of Singapore (NUS), National University
Hospital (NUH), INSEAD Business School, Singapore Institute of
Technology (SIT), and Singapore Science Park.

Notable high-tech and multinational companies such as Google's
Asia-Pacific, Asia-Pacific Economic Cooperation (APEC), Grab,
Razer, and Shopee have situated their offices and headquarters at
one-north.

The one-north precinct consists of several developments such
as Biopolis, Fusionopolis, Mediapolis, Vista, LaunchPad@ one-
north, Rochester Park, Wessex Estate, and Pixel (Ting, 2016). The
16-hectare one-north community park acts as the green link connecting the precincts and serves as a relaxing space for events and
gatherings that promote interaction within the community.

Biopolis is the centre for research and development of biomedical sciences in Singapore. The hub houses many top biomedical
companies and provides space for biomedical research and development activities for leading public and private research institutions,
and promotes peer review and collaboration among the scientific
community.

Fusionopolis is the research and development complex located
at one-north business park. It was designed to foster the growth of
information and communications technology (ICT), media, physical
sciences, and engineering industries. The precinct is an integration
of research organisations and frontier research facilities conveniently and strategically located close to public transport (MRT
station), retail, and accommodation, making it a favourable choice
for companies seeking end-to-end business activities.

Mediapolis is vital to Singapore's Infocomm and media ecosystem.
Mediacorp, the national public broadcaster for radio, television, and
digital content, is housed in Mediapolis alongside the first privately
developed, green, and smart business park ALICE@MEDIAPOLIS
which offers start-up spaces, shared media facilities, flexible work–
live–play–learn spaces, a childcare centre, and retail outlets.

Vista is the precinct designed for commercial and lifestyle purposes at one-north. It acts as the corporate and business support

cluster and houses offices to global companies like Shell and P&G, business hotels, mixed-use complexes (The Star), and residential developments that are located right next to the Buona Vista MRT station.

JTC LaunchPad @ one-north is a high-tech hub that is one of the world's most tightly packed entrepreneurial ecosystems (*The Economist*, 2014) and plays an important role in Singapore's technology start-up landscape. It is situated directly across Fusionopolis and houses digital start-ups, venture capital firms, and tech incubators within the vicinity.

Rochester Park is one-north's latest dining and lifestyle destination. This heritage area consisting of colonial black-and-white bungalows has been given a new lease of life and is home to restaurants and chill-out places.

Built in the 1940s, the Wessex estate (nestled amid lush greenery off Portsdown Road) provides a housing option for one-north's working population.

Pixel, launched by Singapore's Infocomm and Media Development Authority (IMDA), caters to organisations in the Infocomm and media start-up ecosystem. The Pixel space provides the community with access to a wide range of facilities and equipment such as co-working spaces, augmented reality and virtual reality labs, 5G test beds, usability testing labs, prototyping spaces, and sets for filming and media production.

A recent drive and observational walk through one-north and some of the complexes suggests that planners and architects have done a good job in designing futuristic looking buildings and attractive places to eat and meet. Fusionopolis One, for example, is a prime example of JTC's 'mission to develop industrial infrastructure that supports the growth of new industries and transforms enterprises in Singapore' (https://www.jtc.gov.sg/about-jtc/who-we-are).

While exploring the impressive lobby with its tree-like columns (see Figure 7), we also spotted several visionary JTC posters with various messages: "A collaborative space for innovation"; "A meeting point for creative minds"; "Tranquil residential spaces set away from

Figure 6. Covered Skybridge linking Fusionopolis One and Two (Photo credit: Thomas Menkhoff)

the hustle and bustle"; "A variety of recreational and dining options." There is proximity between residential buildings, the various one-north developments, and amenities such as restaurants and retail shops, which suggests that the congregation ambition of planners has been achieved. As shown in Figure 6, there is a covered sky-bridge that links Fusionopolis One and Two.

Figure 8 shows Singapore's largest mural artwork for lyf one-north, the first dedicated co-living property in Singapore. It was created by Columbian artist Didier 'Jaba' Mathieu who painted the gigantic artwork by himself with the help of a jury-rigged gondola. According to Jaba, 'the mural depicts an aspiration for a utopian planet in the future where nature is interwoven in a sustainable city filled with greenery and biophilic designs, connected by a highly sophisticated transport system. This is reflective of the dynamic and innovative spirit of both the one-north district and lyf one-north Singapore' (Popspoken, n.d.). To what extent such murals, linkways, and skybridges do indeed serve as 'connecting nodes to enliven

Figure 7.   Lobby of Fusionopolis One (Photo credit: Thomas Menkhoff)

communities, individuals and experiences' has yet to be ascertained through empirical research.

4.4. *Bras Basah.Bugis (BBB) District — Making It 'Lively' with a 'Vibrant' Street Life*

The Bras Basah.Bugis (BBB) district comprises part of the Rochor and Museum Planning Area. It is located close to the premier Orchard Road shopping belt and the Civic District. The Museum Planning Area (located in the Central Area of the Central Region of Singapore)

Figure 8.   Mural Artwork at lyf One-north (Photo credit: Thomas Menkhoff)

plays a 'bridging role' between the Orchard area and the Downtown Core. Besides a well-developed transport network comprising Stamford Road and Bras Basah Road, Bras Basah MRT/LRT Station, and pedestrian links, it features a 'green lung' in the form of parks and open spaces. Several national monuments are located within the Museum Planning Area such as the Armenian Church, the Cathedral of the Good Shepherd, the Central Fire Station, the Old Tao Nan School (Peranakan Museum), and the National Museum of Singapore.

The URA considers the area to be an 'institutional hub' (11% of the land is set aside for institutional use with reserve sites to be safeguarded for future institutions) due to the many museums here such as the Asian Civilisations Museum and the National Archives of Singapore. The area has a strong arts scene and is home to various places of worship such as churches, mosques, Hindu temples, and synagogues (https://en.wikipedia.org/wiki/Museum_Planning_Area).

BBB is envisioned to be a lively arts, learning, and heritage enclave (CLC, n.d.). Arts, cultural, and educational facilities include the Singapore Art Museum (see Figure 9), Waterloo Street Arts Belt, LaSalle College of the Arts, Nanyang Academy of Fine Arts, the School of the Arts (SOTA), and Singapore Management University (SMU).

The URA has published detailed urban planning parameters and design guidelines for BBB to facilitate innovative designs, as well as "guide the physical development of the area to ensure that individual buildings contribute to, and strengthen the planning vision for the area, while retaining BBB's eclectic character … Gazetted

Figure 9. Singapore Art Museum (Photo credit: Thomas Menkhoff)

monuments and conserved buildings are subject to specific preservation and conservation guidelines respectively, which will take precedence over the guidelines below."

> The term Bras Basah means 'wet rice' in Malay (*beras* means harvested rice with husk removed, and basah means wet) and historically points to a small freshwater stream or the Bras Basah River that used to exist here. In the early days, the area was known for rice trading. The rice would be dried on the riverbanks, but often it would get wet at high tides. Over the years, the area was filled in, and the river Bras Basah (now the Stamford Canal) mostly covered up. The Bugis area is named after the Bugis, a Malay subgroup with origins in South Sulawesi (Celebes) in Indonesia. The Buginese people were once well known for their seafaring and warring skills.

4.4.1. *An expanded arts and cultural precinct*

The URA continues to transform Bras Basah.Bugis, Fort Canning, and the Civic District with its arts and cultural offerings into a vibrant and walkable arts and cultural precinct by creating more park spaces, events, and streetscape improvements. Armenian Street (see Figure 10) has been pedestrianised and a part of it has been converted to an urban botanic garden. It is hoped that road-calming measures and wider sidewalks with student-designed street furniture will motivate more people to walk, socialise, and linger in the precinct: "The treasured historic districts of Little India, Kampong Glam and Chinatown will retain their rich heritage and become more vibrant, through continued place-making efforts in collaboration with the local community and businesses" (URA, n.d.).

Connecting the various BBB stakeholders to nurture 'a lively, arts, learning and heritage enclave' remains work in progress. While there are many noteworthy project initiatives, such as the launch of an arts management major by SMU to leverage its location (SMU is the only local university in Singapore's civic district) or the new University of the Arts Singapore (UAS) which will welcome its first intake in 2024, the diverse complexity of the district requires careful

Figure 10.   Armenian Street next to SMU School of Law (Photo credit: Thomas Menkhoff)

planning and decisive collaborative leadership to harness the area's creative and artistic energy.

In April 2022, Lazada relocated its regional headquarter to Lazada One located within the central district of Bras Basah (see Figure 11). The new office spans 109,000 sq.ft across four floors, together with a visitor centre located on the ground floor: "Featuring open space desks, pantries, mental wellness rooms and soundproof pods, it is designed to facilitate collaborations both within the building and across its regional offices" (Lau, 2022). Other tenants include Zalora, Spotify, as well as Lazada's parent company, Alibaba Group. Whether the ecommerce company will forge an alliance with SMU's School of Computing and Information Systems or the Lee Kong Chian School of Business remains to be seen. Collaborative synergies exist but they need to be harnessed.

The same applies to a possible collaboration between SMU and the National Volunteer & Philanthropy Centre (NVPC), which is

Figure 11. Lazada's Regional Headquarter Opposite SMU (Photo credit: Thomas Menkhoff)

located at The Central, a 25-storey office tower at Eu Tong Street (the distance between NVPC and SMU is about 1km). NVPC is the steward of the City of Good vision for Singapore.

5. Conclusion

In this chapter, we have described how government-led urban planning efforts are directed at Singapore's urban districts, aimed at turning them into 'vibrant,' 'lively,' 'innovative,' 'walkable,' and 'connected' spaces for the benefit of both commerce and communities. Despite challenges in terms of space constraints, the visionary endeavours of Singapore's agencies such as the URA and JTC are noteworthy in terms of their aspirations and decisive implementation. Some of the challenges ahead include obtaining buy-in from all stakeholders to participate in placemaking efforts, harnessing the diverse innovation potential of established districts, and planned mixed-use developments as well the greening of these sites in line with SDG goals.

Geographical proximity alone is not the secret remedy for collaborative innovation to take place. Drawing strength from the strategic interplay between government, academia, industry, research labs, and city communities requires effective knowledge management, participation through informal structures, networking, multi-sector taskforces, conferences, academic publications, etc., and an innovative ecosystem.

Urban skybridges (an elevated type of pedway connecting city buildings in an urban area) or pedestrianised precincts alone are insufficient to create buzz and to make collaborative innovation work. What is important is the existence of trusted networks connecting urbanites and stakeholders from different organisations and their passionate motivation to solve issues together. Combining know-how from different stakeholders as a collective and collaborative cross-boundary learning process with a sense of shared purpose and responsibility is indispensable to make value-creating urban innovation work.

Mixed-use developments are critical in transforming urban centres such as Singapore's central business districts into more 'vibrant' districts by ensuring a mix of commercial, residential, hotel, and entertainment uses. Besides recouping some of the infrastructure investment costs derived from developing urban areas, mixed uses can help to overcome the 'separation of workplace, residence and familial kinship groups' which has caused urbanites in other cities to maintain weak ties in multiple community groups together with high rates of residential mobility (Wellman, 1979).

To avoid the gentrification (Smith, 1996; Harvey, 1989) of prime locations so that they do not become the exclusive turf of wealthy residents and to keep housing accessible, the Prime Location Public Housing (PLH) model was launched by Singapore's authorities in 2021 (*Straits Times*, 2022). Owners of Build-to-Order (BTO) flats in prime locations are required to physically occupy the property for 10 years (the so-called minimum occupation period or MOP) before they can rent or sell the unit.

How Singapore continues to find a balance between planned and 'unplanned' city developments that would allow the intellectual capital of cities to flourish in all dimensions makes for an exciting case. In 'The Art of City Making,' Landry (2006) has discussed the skills required as foundational for this art which goes beyond conventional city planning approaches in architecture, engineering, and land use planning. 'Re-enchanting the city can help cities to become places of solidarity where the relations between the individual, the group, outsiders to the city and the planet are in better alignment.'

Cities can be labyrinths of streets, agglomerations of buildings, and mazes of relations. The urban panorama is a system of close-knit connections between material objects and immaterial factors produced by man. Flexibility in city planning and re-urbanisation approaches allows a different response to the changing needs of a place. Flexibility is intended as the ease with which a system or components of it can be modified and adapted for use in different applications or settings in addition to the ones for which they were originally designed, for example, the use of former industrial spaces for creative industries or the current discussion about turning inner-city shopping areas into housing or non-commercial co-working places. Will some of the planned innovation districts in Singapore be 'rewritten' by its future inhabitants and users?

Different views on urban design processes often lead to confrontation between traditional city planners and planners with a knowledge-based development perspective, as the former often base their planning on maximising the financial value of land use, whereas the latter include the intangible value of city regeneration, which includes understanding the value of local characteristics, culture, and existing social networks. While pragmatism might still be key to Singapore's urban transformation, it can be tinkered with some re-imagination and enchantment by examining the ecosystem of creative space and integrating it with the knowledge networks and transactions prevailing in the four collaborative innovation spaces outlined in this chapter.

References

Alonso, W. (1968). *Location and Land Use: Toward a General Theory of Land Rent.* Cambridge, MA: Harvard University Press.

Biopolis. https://www.worldscientific.com/page/biopolis; https://research.a-star.edu.sg/articles/features/biopolis-ten-years-on/; https://www.asiabiotech.com/09/0924/1320_1326.pdf.

Boschma, R. (2005). Proximity and innovation: A critical assessment, regional studies, 39(1), 61–74.

Burgess, E. W. (1924). *The Growth of the City: An Introduction to a Research Project.* Publications of The American Sociological Association, Vol. XVIII, pp. 85–97.

Chaiechi, T., Wong, C. & Tavares, S. (2020). Urban design and economic growth: An analytical tale of two tropical cities. *ETropic: Electronic Journal of Studies in the Tropics,* 19(2), 172–200. https://doi.org/10.25120/etropic.19.2.2020.3741.

CLC (Centre for Liveable Cities Singapore). (n.d.). Case Study Singapore | Bras Basah.Bugis — Breathing Life Back into BBB. https://www.clc.gov.sg/docs/default-source/urban-solutions/urb-sol-iss-11-pdfs/case_study-sg-bras-basah-bugis.pdf.

Davie, S. (2022). The university and the city. *The Straits Times,* 24 January, https://www.straitstimes.com/singapore/parenting-education/the-university-and-the-city-0.

Fusionopolis. https://www.jtc.gov.sg/find-space/fusionopolis-one.

Herris, C. D. & Ullman, E. L. (1945). The nature of cities. *The Annals of the American Academy of Political and Social Science,* 242, 7–17. DOI: 10.1177/000271624524200103.

Harrison, B. (1992). Industrial districts: Old wine in new bottles? *Regional Studies,* 26(5), 469–483. DOI: 10.1080/00343409212331347121.

Harvey, D. (1989). From managerialism to entrepreneurialism: The transformation in urban governance in late capitalism. *Geografiska Annaler: Series B, Human Geography,* 71, 3–17.

Howkins, J. (2009). *Creative Ecologies: Where Thinking is a Proper Job.* Queensland: University of Queensland Press.

Hoyt, H. (1939). The structure and growth of residential neighbourhoods in American cities. Washington DC: Federal Housing Administration.

Hyundai Motor Group Website. (2020, October 13). https://www.hyundaimotorgroup.com/tv/CONT0000000000002776.

IGI Global. (n.d.). What is industrial district. https://www.igi-global.com/dictionary/networks-industrial-clusters/14207.

IPOS (n.d.). Singapore's global innovation ranking. https://www.ipos.gov. sg/resources/singapore-ip-ranking.

JTC Corp. (2018). One-north — A vibrant work-live-play-learn research and business Park. Accessed 19 May 2022. https://www.youtube.com/ watch?v=SuuwnGhQTdk&ab_channel=JTCCorp.

JTC LaunchPad, Block 71. https://www.economist.com/special-report/ 2014/01/16/all-together-now.

JTC. (2021a). Collaborating for Industry 4.0 success — How can we help each other? Published on 22 November 2021, by Tan Boon Khai, CEO, JTC https://estates.jtc.gov.sg/jid/home/stories/collaborating-for-industry-40-success.

JTC. (2021b). 5 things you should know about Jurong Innovation District — Singapore's advanced manufacturing hub. Published on 13 April, https://estates.jtc.gov.sg/jid/stories/5-things-you-should-know-about-jurong-innovation-district.

JTC. (2021c). JTC sets up one-stop centre, Industry Connect Office @ Jurong Innovation District, Targeting Outreach to 1000 Companies on Digital Transformation Over Next 3 Years, 16 November 2021. https:// www.jtc.gov.sg/about-jtc/news-and-stories/press-releases/jtc-sets-up-one-stop-centre-industry-connect-office-at-jurong-innovation-district.

JTC. (2021d). Open digital platform: The digital backbone of Punggol Digital District, 31 August, https://www.jtc.gov.sg/about-jtc/news-and-stories/feature-stories/open-digital-platform-the-digital-backbone-of-pdd.

JTC. (2022). About JTC: Who we are. https://www.jtc.gov.sg/about-jtc/who-we-are.

Landry, C. (2006). *The Art of City Making*. London-Sterling, VA: Earthscan.

Lau, C. (2022). Lazada relocates regional headquarters to Bras Basah, Youthopia, 12 April. Lazada relocates regional headquarters to Bras Basah, spans four floors with Singapore theme, Youthopia.

Popspoken. (n.d.). Artist Didier 'Jaba' Mathieu brings vibrancy and randomness to Singapore (popspoken.com).

Smith, N. (1996). *The New Urban Frontier: Gentrification and the Revanchist City*. London: Routledge.

Straits Times. (2020). More jobs closer to smarter homes in Punggol, says Teo Chee Hean at ground-breaking of digital district, 17 January. https://www.straitstimes.com/singapore/more-jobs-closer-to-smarter-homes-in-punggol-says-teo-chee-hean-at-groundbreaking-of.

Straits Times. (2021a). 4 international companies to set up base in Punggol Digital District, creating about 2,000 jobs, 28 July. https://www.

straitstimes.com/business/economy/4-international-companies-to-set-up-base-in-singapores-punggol-digital-district.

Straits Times. (2021b). 10-year plan for Singapore manufacturing to grow 50% by 2030: Chan Chun Sing, 25 January 2021. https://www.straitstimes.com/business/economy/10-year-plan-for-singapore-manufacturing-to-grow-50-by-2030-chan-chun-sing.

Straits Times. (2022). Singapore risks gentrification if public housing left to market forces: Desmond Lee, 22 September. https://www.channelnewsasia.com/singapore/ministry-national-development-desmond-lee-forward-singapore-session-gentrification-public-housing-free-market-sources-2964136.

The Economist. (2014). All together now — What entrepreneurial ecosystems need to flourish: All together now. *The Economist,* January 16.

Ting, I. (2016). Straits Times. Pixel Studios, IMDA's space for digital content creators, opens at one-north. Accessed 19 May 2022. https://www.straitstimes.com/tech/pixel-studios-imdas-space-for-digital-content-creators-opens-at-one-north-today; https://www.imda.gov.sg/programme-listing/pixel.

URA. (n.d.). Pungol Digital District, https://www.ura.gov.sg/Corporate/Guidelines/Urban-Design/Punggol-Digital-District; https://en.wikipedia.org/wiki/One-north.

URA. (n.d.a). Strengthening Economic Gateways, https://www.ura.gov.sg/Corporate/Planning/Master-Plan/Themes/Local-Hubs-Global-Gateways/Strengthening-Economic-Gateways.

URA. (n.d.b). Celebrating our Arts, Culture and Heritage, https://www.ura.gov.sg/Corporate/Planning/Master-Plan/Regional-Highlights/Central-Area/Celebrate-Arts-Culture-Heritage

Wellman, B. (1979). The community question: The intimate networks of East Yorkers. *American Journal of Sociology,* 84(5), 1201–31.

https://doi.org/10.1142/9789811293108_0002

Chapter 2

Planning Ideas for an Educational Living Energy Lab — A Student Project Assignment in a Smart City Course

Nicole Bremstaller, Deng (Frank) Xingyuan, Kamille Storm Kastrup, Annika See, and Irene Zhang

1. Introduction

Upon realising the consequences brought about by environmental damage, people around the world are expressing their concerns about the future risks of the energy crisis. According to the International Energy Association (IEA), in addition to the shortage of fuel, inflation, and poor economic conditions, people are facing the dilemma of unaffordable energy. As IEA Executive Director Fatih Birol has stressed, "The world has never witnessed such a major energy crisis in both depth and complexity" (Stringer, 2022).

In the first half of 2022, the average price of natural gas in Europe was four times higher than the same period in 2021. The price of coal was more than three times that at the same period in 2021, resulting in the tripling of electricity prices in massive markets compared with the same period the previous year (IEA, 2022). However, the high cost of electricity did not impact the increasing demand for energy because coal is still the most frequently used energy source in many countries around the world. The emission of carbon is therefore affecting the environment and exponentially accelerating climate change.

The world is concerned about tackling an energy transition challenge from fossil fuels to renewable energy sources, with many countries setting goals to reach carbon neutrality by 2050 (Li *et al.*, 2022). Solar energy is one of the most promising renewable energy sources, especially in warmer regions. Compared to other alternative sources of renewable energy, what makes solar energy particularly promising is the fact that the total annual solar radiation on Earth is about 7500 times greater than the world's total amount of annual primary energy consumption (Li *et al.*, 2022).

For our team's final project in the Singapore Management University summer school course "Innovations for Asia's Smart Cities," we chose to tackle the topic of "Public–Private Partnerships (PPPs): Energy (Carbon) Crisis and Living Energy Labs." To address the energy crisis, our team's goal was to create a proposal that leverages PPPs to create a "living energy lab" plan. We decided to focus specifically on researching solar energy as a source of renewable energy. Role playing as a team employed by the university, our project proposed a strategy and an implementation plan to create SMU's very own living energy lab.

The term "living lab" refers to a place-based research platform that typically leverages knowledge clusters — such as college campuses — as a test bed for innovation and the co-production of sustainability knowledge (MIT a/b, n.d.). The living lab concept points to a new research and development environment dedicated to cultivating user-centred and future-oriented scientific and technological innovation modes. Living labs use scientific research institutions as a link to establish an open and more innovative society. Besides institutions of higher learning, other key stakeholders include the government as well as enterprises and their networks.

To make better use of resources beyond the government sector, several living labs around the world have adopted a collaboration method called "Public–Private Partnership" (PPP). This type of part-nership enables the private sector to work alongside the public sector to provide goods and services to the public. PPPs can encourage innovation in the areas of public governance and private investment.

2. Solar Energy

Solar energy is energy collected from sun rays. It can be harvested either as heat (i.e., Solar Thermal Energy Systems) or electricity (i.e., Photovoltaic Systems), or both electricity and heat simultaneously (Li *et al.*, 2022). Other related technologies include solar cooling, natural lighting, and solar fuel (United Nations, 2022). It is important to note that solar energy is of intermittent nature and highly dependent on environmental factors such as clouds, shadow, weather conditions, and, obviously, the amount of sunlight. There may therefore be imbalances between supply and demand in certain regions or at certain times. Currently, there is an established market for solar energy systems established by policies, investments, and research funding from governmental and non-governmental organisations (Kabir *et al.*, 2018).

2.1. *Benefits of Solar Energy*

The use of solar energy brings numerous benefits, the most important one being the sustainable and renewable nature of energy conversion and collection. The abundant availability of sunlight to supply solar energy is also very convenient (Li *et al.*, 2022). Moreover, the use of solar energy leads to less carbon emissions and reduces the need for importing fuels, which aids environmental sustainability and increases energy security (Government of Singapore, 2022b). Furthermore, the deployment of solar energy systems will increase and improve job opportunities, stimulating local economies where these systems are implemented. Over the past few years, a lot of research has been done to increase efficiency resulting in decreasing costs of solar energy systems. Solar energy systems are therefore becoming increasingly affordable. Another positive aspect of this renewable energy source is the flexibility of installation, as solar energy systems can be installed on rooftops, windows, and even on platforms floating in the ocean (Kabir *et al.*, 2018).

2.2. *Challenges of Solar Energy*

Although it may appear like the problems outweigh the benefits of solar energy, it is important to bear in mind that researchers are working on finding novel solutions to existing problems. The biggest challenge in the sector is finding appropriate technologies to harvest solar energy.

The first significant hurdle encountered when utilising solar energy is the variation of solar radiation throughout the year, e.g., different amounts of sunlight in summer vs. winter. This challenge is a factor currently outside of our control and cannot be solved with technology. Another issue is the storage of solar energy as demand and supply are not always equal (Li *et al.*, 2022). While there are already several storage systems available, they are still expensive, and more research and development work is necessary to create such systems for higher capacities (Government of Singapore, 2022d). Other obstacles include inadequate infrastructure, insufficiently skilled manpower, and high investment costs (Kabir *et al.*, 2018).

Finally, the issue of the recyclability of solar panels needs to be addressed. Solar panels only have a lifespan of approximately 30 years, depending on the type of manufacturing material (United Nations, 2022). By 2050, around seventy million metric tons of solar panels will be at the end of their lifespan, and current recycling solutions are insufficient. By disposing of solar panels inappropriately, valuable resources are wasted and toxic materials used in panels could leak, consequently polluting land. Some initiatives to start recycling solar panels already exist. For example, the non-profit organisation PV CYCLE works with the European Union to acquire used solar panels and recycle them. According to European Law, solar panels must be recycled correctly. In the US, where solar energy usage is high, no such laws exist yet (Bates, 2020).

2.3. *Global Solar Energy Research*

Many different countries have invested in researching and implementing solar energy systems. The countries with the highest usage of solar energy are China, United States, Japan, Germany, and India. China has the world's largest solar energy deployment. Japan has been looking into alternative energy sources since the Fukushima disaster in 2011. In addition to the Fraunhofer Institute for Solar Energy Systems ISE, which has 1400 employees, Germany also has many small-scale private installations (NS Energy Staff Writer, 2021). Other regions in the world that are increasing investments in solar energy include Africa, Latin America, and the Middle East.

The biggest potential lies in Africa due to its high solar radiation and abundance of space. However, the world's highest solar radiation per square meter of all continents is in Australia, and thus the country has the best solar energy resources in the world. Comparable levels of radiation are only found in Africa, Southwestern USA, Mexico, and South America (Bennington-Castro, 2019).

3. Towards a Living Energy Lab Solution for SMU

3.1. *Elements of a Living Lab*

Living labs act as intermediaries between citizens, research organisations, companies, and cities to co-create innovations that enhance the quality of life within their region (U4IoT, 2017). While each living lab has different priorities, they all share core elements that make up the backbone of the lab, namely, a multi-method approach, user engagement, multi-stakeholder engagement, real-life setting, and co-creation. These are described in Table 1.

3.2. *Living Lab Objectives*

To successfully develop innovations that connect with users and stakeholders, we identified three key objectives for the lab, namely, exploration, experimentation, and evaluation.

Table 1.   Core Elements of a Living Lab

Multi-Method Approach	There is no single methodology for creating a living lab; it requires the use of multiple methods to find the best fit for the lab's purpose.
User Engagement	The purpose of the lab is to create innovations for the user. Therefore, the lab is responsible for creating activities that involve users to gain success.
Multi-stakeholder Participation	To continuously evolve and gain new perspectives, the input of representatives from private and public sectors, academia, and people will be beneficial.
Real-Life Setting	The purpose of the living lab is to imitate real-life current and future states to evaluate the feasibility of innovations.
Co-Creation	Living labs strive to create opportunities for all stakeholders to actively engage in the process from beginning to end.

Source: U4IoT (2017).

The process of **exploration** is also known as the "problem–solution fit" because the main purpose of this stage is to identify elements of the problem and understand the current state (U4IoT, 2017). Analysing the current state involves identifying current trends, habits, and practices of the target users. Using this process, we can identify specific problems that must be addressed while taking into consideration the context of the problems. After this step, uncovering the needs and wants of the user will allow us to identify opportunities for improvement and solutions. This will trigger the brainstorming of future states that can potentially be pursued. Overall, the exploration stage enables a better understanding of the users and their context, which can potentially inspire ideas for innovation that can become solutions for users.

Following the exploration phase, we enter the **experimentation** phase where prototypes are created in response to the future states developed during the exploration stage. During the experimentation stage, an element of "real-life setting" will be used depending on the type of prototype created. The main activity here is to evaluate how things will work in the future. These tests can be short or long, and they can come in different forms as they should be designed to be compatible with the target user. In summary, the experimentation stage materialises solutions created based on predicted future states and tells us whether to go back to the first stage or move into the evaluation stage.

The last stage is to **evaluate** the innovation created in response to the current state which illustrates the potential impacts and value added by the innovation. This stage is also known as "product–market fit" as it evaluates the value and advantages of the innovation in the face of the user market. While substantial research and testing are necessary, elements such as pricing, advertisement, location, and accessibility must also be considered before the final product is released to the public.

4. Examples of Living Labs

To develop a successful and relevant living lab, our team conducted research on educational institutions around the world that already possess well-functioning living labs. By researching existing case studies, summarised in Table 2, we gained inspiration for key features to be included in the living lab proposal.

Table 2.   Summary of Living Lab Case Studies

Name of Living Lab and Location	Key Features	Key Learnings
The Cambridge Green Challenge Living Laboratory — *University of Cambridge, England* (University of Cambridge, 2022).	Focuses on making students work with sustainability through projects, internships, and research.	Experienced professional resources help ensure the success of the living lab.
	Helps students coordinate, finance, and run projects that have an impact both inside and outside the university.	An opportunity to work with actual cases makes the lab more relevant.
	Offers paid internships with the opportunity to collaborate with professionals in the field and on real cases.	

(*Continued*)

Table 2. (*Continued*)

Name of Living Lab and Location	Key Features	Key Learnings
Architectural Lighting Lab — *Royal Danish Academy, Denmark* (Royal Danish Academy, n.d.).	Includes various opportunities for testing natural and artificial light.	Physical models are an effective way to test large-scale projects and help concretise major issues.
	Possesses a physical model workshop and a machine that functions as an artificial sun that can be set to any time and date.	Opening the lab to stakeholders other than students increases the relevance and broadens the uses of the lab.
	Companies and other stakeholders can rent and access the laboratory and a consultant.	
Five Living Labs, Office of Sustainability — *Massachusetts Institute of Technology (MIT)* (Massachusetts Institute of Technology 2022a/b).	The Office of Sustainability at MIT facilitates five different living labs, all with a focus on specific sustainability issues and each possessing their own area of expertise.	Using the entire campus area as a larger-scale model for solutions makes it possible to test projects that are not convertible to model size.
	Uses the entire campus space as a model for testing sustainable projects and as a local test bed for global solutions.	

5. Anticipated Outcomes of the Proposed SMU Living Lab

The proposed SMU Living Lab shall focus on innovations that align well with established Smart City frameworks (e.g., the EU Smart Cities Framework which we used for our project). As Singapore continues to develop SMART ways to grow the country, we believe that

Table 3. EU Smart Cities Framework[*] Applied to the Living Lab

Smart City Component	SMU Living Lab Outcomes
Smart Living	The living lab serves as an educational facility for the public to engage with.
Smart Mobility	Its downtown location increases the living lab's accessibility to all communities.
Smart Economy	The living lab promotes an innovative spirit through its iterative nature and the PPP structure enlarges the productivity of the lab.
Smart People	The public will learn about sustainable energy innovations through the living lab, and stakeholders from both the public and private sectors will be engaged in the living lab.
Smart Governance	With the help of the Singapore government and the input of stakeholders, the living lab concept can be scaled up and replicated.
Smart Environment	Environmental protection and sustainable resource management are achieved through solar energy R&D.

[*]https://www.smart-cities.eu/

it is important for the lab to be aligned with Singapore's Smart Nation aspirations and community needs.

By 2030, the Singapore government aims to supply around 350,000 households a year with solar energy (Government of Singapore, 2022b). Currently, Singapore can generate 1580 kilowatt-hours per square meter per year with the received solar irradiance. Although solar energy is a very promising renewable energy source for the island nation, one of the main problems impeding the implementation of solar energy systems is the high population density and land scarcity, which limit the amount of space to mount systems (Government of Singapore, 2022b).

Interesting research is being conducted into how we can decarbonate the world through renewable sources of energy. With the solar energy sector booming, research on solar energy innovations is in high demand. Current innovation efforts include solar thermochemistry, solar cooling, "floatovoltaics," and solar-driven water distillation (Li *et al.*, 2022).

6. Leveraging Public–Private Partnerships (PPPs) for Multi-Stakeholder Participation

The success of a living lab depends on ensuring that every stakeholder is fully invested in the initiative and is willing to contribute to the project. Hence, the evaluation and selection of stakeholders from both the public and private sectors is a vital step in the implementation of a living lab. To fully leverage the Public–Private Partnership (PPP) framework for the proposed living lab, potential stakeholders will have to be carefully evaluated and selected using the main criteria to see how their strategic goals are aligned with that of the lab. In the context of our SMU living lab project, the relevant stakeholders include SMU (relevant administrative units, students, and faculty), the Singapore Government, and solar energy companies. These are summarised in Table 4.

Table 4. Summary of PPP Stakeholders and Goal Alignment

Contributors/ Strategic Goals	SMU	Singapore Government	Solar Energy Companies
Stakeholders	Office of Facilities Management and Planning; Students; and Faculty.	Energy Market Authority (EMA).	Singapore Power; Union Power; LYS Energy Group; SunSeap; and Cleantech Solar.
Goal Alignment	SMU Vision 2025 prioritises sustainable living. The university employs strategies of transformative education, conducts cutting-edge research, and aspires to be an engaged city university.	EMA follows the 4Rs for promoting sustainable energy: right pricing, regulation reduction, raising demand, and R&D; benefits from the sale of excess solar energy back to the grid; and aligns with Smart City components.	Aim to hire local talent and gain access to research and development opportunities and funding.

6.1. *SMU*

The SMU living lab is a project that would be executed through a partnership with the SMU Office of Facilities Management and Planning. With the mandate of managing campus infrastructure and spearheading projects for the innovation of space, this office would oversee the project. SMU students can be hired for executing day-to-day operations, and professors or faculty members can assist in strategising and managing the lab. As a stakeholder, SMU provides the talent pool and knowledge cluster required for maintaining the living lab.

An obvious, low-hanging fruit project would be to install Electric Vehicle (EV) charging stations in the two SMU car parks with the electricity generated using solar energy systems installed on the rooftops of SMU buildings.[1] The initial phase will involve installing three EV charging stations at the Administration Building car park and seven stations at the main SMU car park. The project will include the purchase and installation of solar systems equipment and EV charging stations; installation of cables and other infrastructure; and systems integration forming the "glue" to ensure that all these subsystems are able to work together seamlessly. Due to the multi-disciplinary nature of this project, industry partners with the appropriate experience will be invited to participate in the project. For cases where funding is required, the budget will come from the combined PPP fund.

At the strategic level, the implementation of a living lab at SMU aligns with SMU's Vision 2025, a strategic plan which prioritises achieving sustainable living by employing strategies of transformative education, cutting-edge research, and creating an engaged city university (Singapore Management University, 2022). We selected SMU as the main educational institution for implementing the living lab over other universities with a stronger area of technological expertise (e.g., Nanyang Technical University and National University

---

[1]In 9/2022, SMU launched its "Sustainability Blueprint" with four key strategies: (i) Cultivate a Greener University; (ii) Develop Change Agents through Sustainability Education; (iii) Drive Impactful Research; and (iv) Foster Resilient Communities. The blueprint is aligned with the 17 UN Sustainability Development Goals (SDGs).

of Singapore) due to its ideal location in downtown Singapore with its proximity to relevant partner companies, accessibility, and its deep entrepreneurial and management knowledge. Using SMU as a testing ground will facilitate the replication and scaling of the living lab concept across other higher educational institutions in Singapore.

6.2. *Singapore Government*

The other stakeholder of this PPP is the Singapore government, and more specifically, the Energy Market Authority (EMA). Currently, the Singapore government has implemented incentive schemes to motivate people to install solar panels on their rooftops, with the option to sell excess solar electricity to the electrical grid (Government of Singapore, 2022c). This initiative should be continued and additional support from the government in the form of funding, subsidies, and official endorsements would also aid the implementation of the living lab. EMA's involvement in the living lab would align with its goals for promoting sustainable energy, or the four "Rs" — "Right Pricing," "Regulation Reduction," "Raising Demand," and "Research and Development" (Government of Singapore, 2022a). The living lab also aligns with Smart City initiatives, which are supported by the Government of Singapore.

6.3. *Solar Energy Companies*

Corporate partners of this PPP include local Singaporean firms and start-ups such as Singapore Power, Union Power, LYS Energy Group, SunSeap, and Cleantech Solar. In the proposed living lab project, the private sector would spearhead the research and development (R&D) of solar energy technologies, provide critical resources or assets for R&D, and collaborate with SMU stakeholders to bring new ventures in solar energy to fruition. For example, SunSeap has a proven track record of being Apple Computer's partner in installing solar energy systems on top of 800 buildings in Singapore (Apple Computer, 2018).

The involvement of solar energy companies in the proposed living lab would be greatly beneficial in terms of hiring local talent and gaining access to R&D opportunities and funding. The living lab gathers students most enthusiastic about solar energy, innovation, and technology in one place, and provides opportunities for companies to directly work with these students, effectively creating a strong pool of candidates for their talent acquisition efforts.

7. Living Lab Implementation

7.1. *Implementation Timeline*

As shown in Figure 1, the implementation of the SMU living lab spans 4 years and consists of three main phases: (1) Planning, (2) Executing, and (3) Monitoring.

During the planning phase, all relevant stakeholders will be consulted to clarify and reaffirm the objectives of the living lab and ensure that their participation in the lab aligns with their strategic goals. After securing funding for the lab and establishing the necessary public and private partnerships, the lab must begin hiring students and staff for operations and create a comprehensive marketing campaign to promote the living lab to the public. During the execution phase, the launch of the living lab will be followed by R&D and execution of solar energy projects. Following the completion of these projects, solar energy education will commence through public

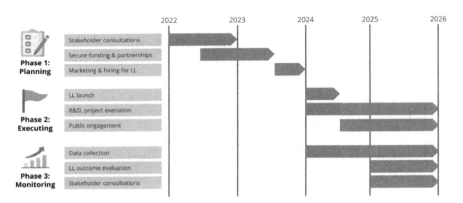

Figure 1. Four-Year Timeline of Implementation for SMU Living Lab

engagement. Data collection will begin with the launch of the living lab and will continue to play an essential role in the final phase of monitoring. Evaluation data will help ascertain to what extent the living lab's outcomes were achieved and provide quantitative evidence for stakeholders with regard to the performance of the lab.

7.2. *Financing the SMU Living Lab*

Ensuring the long-term financial viability of the SMU living lab is critical to achieving and maintaining success over time. Given the nature of PPPs, funding for the living lab must be a mixed balance between the public and private sectors — existing frameworks such as the one shown in Figure 2 can be used as guidance for financing the SMU living lab.

There are several ways to finance the lab for each stakeholder: examples include low-interest public loans and subsidies from the government; government or SMU scholarships for students to participate in the lab; and the provision of internships or career opportunities for students from private firms.

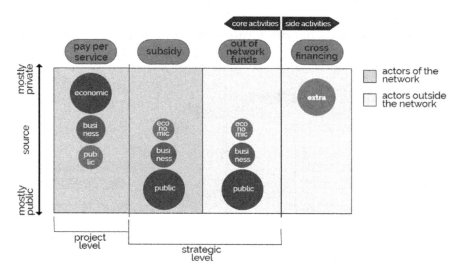

Figure 2.   Funding Mix Framework (Gualandi & Romme, 2019)

8. Conclusion

Through this project, our team gained a comprehensive understanding of the incredible potential of solar energy and the importance of living labs and PPPs in bringing such innovations to life. The class context provided an additional "Smart City" lens that helped us understand and frame this project so that it amounted to more than just a sustainability-related initiative. Studying the core components of a "Smart City" and relevant frameworks enabled us to appreciate the importance of designing an impactful, collaborative living energy lab that can create sustainable impact with a specific application to solar energy. As a team, we are very grateful for the opportunity to participate in this class, and there is no doubt that we will bring our learnings into our future academic and career endeavours.

References

Apple Computer. (2018, April 10). Apple now globally powered by 100 percent renewable energy. https://www.apple.com/sg/newsroom/2018/04/apple-now-globally-powered-by-100-percent-renewable-energy/.

Bates, A. (2020). Solar panels are starting to die. What will we do with the megatons of toxic trash? Grist Magazine, Inc., 13 August. https://grist.org/energy/solar-panels-are-starting-to-die-what-will-we-do-with-the-megatons-of-toxic-trash/.

Bennington-Castro, J. (2019). Floating solar farms: How "floatovoltaics" could provide power without taking up valuable real estate. NBC Universal, 12 February. https://www.nbcnews.com/mach/science/floating-solar-farms-how-floatovoltaics-could-provide-power-without-taking-ncna969091.

Government of Singapore. (2022a, March 14). Overview (Renewable Energy). https://www.ema.gov.sg/Renewable_Energy_Overview.aspx.

Government of Singapore. (2022b, March 14). Solar photovoltaic systems. https://www.ema.gov.sg/Solar_Photovoltaic_Systems.aspx.

Government of Singapore. (2022c, March 14). Guide to solar PV. https://www.ema.gov.sg/Guide_to_Solar_PV.aspx.

Government of Singapore. (2022d, March 14). Energy storage system. https://www.ema.gov.sg/energy-storage-system.aspx.

Gualandi, E. & Romme, A. (2019). How to make living labs more financially sustainable? Case studies in Italy and the Netherlands. *Engineering Management Research*, 8(1), 16. DOI: 10.5539/emr.v8n1p11.

IEA. (2022, July 20). Global electricity demand growth is slowing, weighed down by economic weakness and high prices. IEA. https://www.iea.org/news/global-electricity-demand-growth-is-slowing-weighed-down-by-economic-weakness-and-high-prices.

Kabir, E., Kumar, P., Kumar, S., Adelodun, A. A. & Kim, K. (2018). Solar energy: Potential and future prospects. *Renewable and Sustainable Energy Reviews*, 82 (2018), 894–900. DOI: 10.1016/j.rser.2017.09.094.

Li, G., Li, M., Taylor, R., Hao, Y., Besagni, G. & Markides, C. N. (2022). Solar energy utilisation: Current status and roll-out potential. *Applied Thermal Engineering*, 209, 1–6. DOI: 10.1016/j.applthermaleng.2022.118285.

Massachusetts Institute of Technology. (2022a). Defining living labs. MIT Office of Sustainability. https://sustainability.mit.edu/defining-living-labs.

Massachusetts Institute of Technology. (2022b). Discovering living labs. MIT Office of Sustainability. https://sustainability.mit.edu/living-labs.

NS Energy Staff Writer. (2021, July 13). Top five countries with the largest installed solar power capacity. NS Energy. https://www.nsenergybusiness.com/features/solar-power-countries-installed-capacity/.

Royal Danish Academy. (n.d.). Architectural lighting lab. https://royaldanishacademy.com/workshop/architectural-lighting-lab.

Singapore Management University. (2022). Vision 2025. https://www.smu.edu.sg/about/vision-2025.

Stringer, D. (2022, July 12). Worst of global energy crisis may still be ahead, IEA Says. *Bloomberg*. https://www.bloomberg.com/news/articles/2022-07-12/worst-of-global-energy-crisis-may-be-ahead-iea-s-birol-warns.

United Nations. (2020). What is renewable Energy? https://www.un.org/en/climatechange/what-is-renewable-energy?gclid=Cj0KCQjwzqSWBhDPARIsAK38LY88Z0fFdy5cVBDS4zBqyUbGK0DMEeu1_e9ApaAknag1IuMizDUw-WIaAsBqEALw_wcB.

University of Cambridge. (2022). Living laboratory for sustainability. The University of Cambridge. https://www.environment.admin.cam.ac.uk/living-lab.

U4IoT. (2017). *Living Lab Methodology Handbook.*

Chapter 3

Industry 4.0 Adoption by Small and Medium-Sized Manufacturing Enterprises in Singapore: A Case Study

Surianarayanan Gopalakrishnan and Thomas Menkhoff

1. Introduction

The term 'Industrie 4.0,' also known as I4.0 or I4, originated in Germany and was coined during the 2011 Hannover Fair. Other leading industrial nations have widely recognised this term: the United States calls it the 'Connected Enterprise' and the United Kingdom refers to it as the 'Fourth Industrial Revolution.'

Industry 4.0 adoption is expected to profoundly impact the entire spectrum of industries, especially in manufacturing. By using a confluence of automation, data, and digitalisation, Industry 4.0 aims to radically transform how organisations operate presently while increasing productivity, enhancing flexibility, reducing costs, and improving efficiency. More companies are strategically embracing the so-called Industry 4.0 approaches to leverage opportunities arising from newly connected computers and increasingly autonomous automation systems (e.g., robotics) equipped with intelligent machine learning algorithms that control the robotics without much human input. In these 'smart' factories (Belton *et al.*, 2019), cyber-physical systems (i.e., independently operating systems that self-optimize, communicate with each other, and ultimately optimize production) monitor the physical manufacturing processes and play an increasingly important role in terms of decision-making.

Industry 4.0 signifies three mutually interconnected factors (Zezulka *et al.*, 2016), namely digitisation and integration of any technical-economic networks, digitisation of products and services, and new market models. At the core of this new smart manufacturing paradigm (McKewen, nd) is the Internet of Things that drives the conversion of traditional factories into a 'smart' manufacturing environment called 'Industry 4.0' (Kagermann *et al.*, 2013), resulting in an increasingly intelligent, connected, and autonomous factory with outstanding dynamic capabilities (Teece, 2010; Eisenhardt & Martin, 2000). Smart manufacturing technologies include big data processing, machine learning, advanced robotics, cloud computing, sensor technology, additive manufacturing, and augmented reality. By using predictive big data analytics, deep learning, or sentiment/image analysis, business leaders can identify patterns and trends in vast reams of big data. It allows them to make 'smarter' decisions (e.g., about the loss of customers or the necessary service inspection of equipment) and potentially to become more competitive in real time.

2. Research Goal, Rationale and Questions

2.1. *Conceptual Framework: Sustaining Singapore's Competitiveness in Manufacturing and Technology Innovation through Industry 4.0*

Characteristic features of an Industry 4.0 system (Marr, 2016; Schwab, 2016) include interoperability (machines, devices, sensors, and people that connect and communicate with one another), information transparency (the systems create a virtual copy of the physical world through sensor data to contextualize information), technical assistance (both the ability of the systems to support humans in making decisions and solving problems and the ability to assist humans with tasks that are too difficult or unsafe for humans) and decentralized decision-making (the ability of cyber-physical systems to make simple decisions on their own and become as autonomous as possible).

During the last few years, there has been increased interest in Industry 4.0 initiatives across the Southeast Asian economies (ASEAN), with most governments taking active measures to push toward this implementation. ASEAN aspires to be the world's fourth largest economic block by 2050, and Industry 4.0 is a critical element of this growth vision across the region. Most countries, including Singapore, have created a road map with a vision for Industry 4.0. However, there is arguably limited adoption despite all the initiatives and support extended by the various governments.

The manufacturing sector is critical for Singapore's economy as it makes up about 21% of Singapore's GDP, accounting for 14% of its workforce. Due to the lack of research data on the firms' receptiveness toward advanced Industry 4.0 technologies, it is critical to examine the impact of automation, digitization, and web-based production systems on the business models of local manufacturers in terms of successful 'production optimization,' greater 'product customization,' 'predictive' maintenance and 'visionary' investments into automation solutions. According to the Singapore Business Federation, about 60% of local small and medium-sized enterprises (SMEs) have not made significant technological changes. Possible reasons include resistance to change, lack of technological knowledge and resources, and insufficient leadership foresight. Many local firms are still operating in 'the Industrial 2.0/3.0' era rather than embracing opportunities arising from smart 'Industry 4.0' technologies. In smart factories, cyber-physical systems (i.e., independently operating systems that self-optimize and communicate with each other) monitor the physical manufacturing processes and play an increasingly important role in decision-making. At the core of this new smart manufacturing, the paradigm is IoT, resulting in an increasingly intelligent, connected, and autonomous factory with excellent dynamic capabilities.

Singapore has an adequate level of understanding with companies of the need to migrate to Industry 4.0, and the benefits of implementing this also generate much excitement. However, unlike Germany or the United States, the pace of adoption has been muted beyond a few early adopters. Singapore's Economic Development Board (EDB) launched a Smart Industry Readiness Index (SIRI) in 2017 to help firms measure their progress and understand the gaps in implementation. Technology, Process, and Organisation were the three building blocks of Industry 4.0 identified by EDB. The Model Factory at A*STAR's Advanced Remanufacturing and Technology Centre (Launch of A*STAR's Model Factory) and the Operation and Technology Roadmap (OTR) launch aids firms in realise their transformation on this smart manufacturing journey. However, the take-up rate by companies seems to be still low.

In 2015, the Agency for Science, Technology, and Research (A*STAR) started a Future of Manufacturing (FoM) Initiative in close consultation with the Ministry of Trade and Industry (MTI), the Economic Development Board (EDB), and SPRING Singapore under the Government's Research, Innovation & Enterprise 2020 plan. The deliberations involved extensive engagement with industry and trade associations. The initiative's goal is to sustain Singapore's competitiveness in manufacturing and technology innovation. It is a location of choice for developing, test-bedding, and deploying advanced groundbreaking technologies in the manufacturing sector: "The three key thrusts of A*STAR's FoM Initiative are the public-private partnership platforms of Tech Access, Tech Depot, and Model Factories, that aim to drive technology innovation, knowledge transfer and adoption across the manufacturing industry (A*STAR Science and Engineering Research Council (SERC)."

2.2. *Conceptual Framework: Purpose of Research and Research Question*

There are limited Singapore-related studies to understand the driving forces and barriers to adopting and implementing Industry 4.0. Most studies focus on the technology aspects and provide limited understanding from SME's holistic business model viewpoint. According to the Singapore Business Federation, about 60% of local SMEs have not made significant adjustments to adapt to technological change. We attempt to address this significant knowledge gap and wish to contribute to the minimal Asian management literature on the smart manufacturing technology research gap's impact on understanding the experts across the ecosystem. We believe that this approach will help us better understand current business dynamics, potential issues, focus areas, and developments related to the pace of adopting advanced manufacturing approaches within Singapore's manufacturing sector to help catapult Singapore's manufacturers to the next level. The core research question addressed in this chapter is: "How do Singapore's manufacturing small and medium enterprises embrace Industry 4.0 to enhance their business models?"

3. Research Goal, Rationale and Questions

3.1. *Method*

Our study adopted a dual-phase approach to developing a model that can clarify factors influencing the readiness of SMEs for the adoption of Industry 4.0. The first phase involved expert semi-structured interviews (via Zoom) with eight key decision-makers and heads across government agencies, Institutes of Higher Learnings (IHLs), suppliers/providers of technology, business associations and the local SME sector. A structured questionnaire was devised based on the current body of knowledge in Industry 4.0.

The objective of the expert interviews was to allow us to dive deeper into the issues involved to develop a comprehensive understanding of the challenges faced by SMEs in Singapore. A qualitative approach allows the interviewees to express themselves freely while providing us researchers with the option to seek additional inputs and quality guidance. "Qualitative research is conducted through intense and prolonged contact with a 'field' or life situation" (Miles & Huberman, 1984). Interviews were recorded with the participant's permission and then transcribed verbatim using transcription software, followed by an analysis of emerging themes using software

Table 1. Factors Impacting Industry 4.0 Adoption in SMEs

Sl. No.	Factors
1	Capability and Competency
2	Collaboration — Internal and External
3	Culture of Company
4	Business Model Innovation
5	Skilled Labor Shortage and Dependencies
6	Mindset and Resistance to Change
7	Productivity and Efficiency
8	Return on Investment and Capital Costs
9	Government Subsidies, Incentives and Support
10	Talent Shortage
11	Technological Push by the Government
12	Top Management Support
13	Turnkey Solution Providers
14	Uniform Standards
15	Strategy and Implementation Roadmap

(Nvivo). Based on a literature review and the expert interviews, we eventually arrived at several factors or categories (see Table 1) impacting Industry 4.0 adoption in SMEs.

During phase 2 of the research project, we interviewed SME owner managers (on-site) as part of our case study approach. Qualitative data analysis techniques are well established in the management field, comprising three different procedures: data reduction (e.g., discarding irrelevant data in transcripts), data display (e.g., specific graphical formats derived from the data), and drawing conclusions based on field notes and emerging themes (Miles & Huberman, 1984). In line with the criteria of qualitative research, respondents' transcribed statements were clustered into common (raw data) themes and grouped into the so-called first-order themes. Units with different meanings and general dimensions were established based on second-order themes (Eisenhardt & Graebner,

2007). Table 1 summarizes the key outcomes of this step as far as the expert interviews are concerned.

4. Findings of Expert Interviews

The expert interviews helped in identifying five key drivers and four main barriers regarding the adoption and implementation of Industry 4.0 approaches, as illustrated in Figure 1.

Key drivers include technological push and incentives by the government, skilled labour shortages, productivity and efficiency gains, the pressure to innovate business models (Girotra & Netessine, 2014; Osterwalder *et al.*, 2010; Taran *et al.*, 2019) due to increased competition and the impact of COVID-19. Barriers to adoption revolve around return-on-investment issues, capability constraints, lack of ecosystem, and mindset factors. All experts agreed that Industry 4.0 was well recognised across the manufacturing SMEs in Singapore due to the top-down kind of initiatives embarked upon by the government in Singapore and well supported by the agencies supporting this initiative.

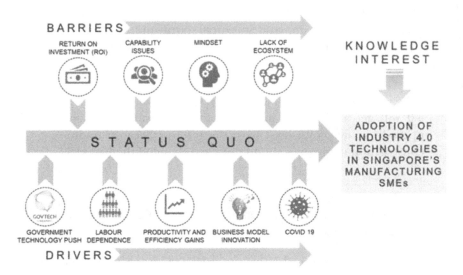

Figure 1. Drivers and Barriers of Industry 4.0 Adoption and Implementation

"This is because Singapore has identified Industry 4.0 as an important skill and an important capability to improve productivity and to keep manufacturing content as a percentage of GDP existing in Singapore, and of course there has been a lot of downward pressure on that over the last few years."

(Interviewee 2)

The expert views varied in terms of readiness to adopt and implement Industry 4.0 and SMEs' capabilities. They ranged between 5 and 7 on a scale of 0 to 10 (0 = very low; 10 = very high) to measure SMEs' readiness and capabilities to adopt and implement Industry 4.0 initiatives. The size and sector in which the SME operated played a crucial role in adopting Industry 4.0. It was unanimously felt that larger SMEs (USD 100 million and above) were very well positioned to embrace Industry 4.0. In contrast, medium-sized SMEs (USD 10–100 million) were now more willing to embrace Industry 4.0, especially on the back of COVID-19 bottlenecks and issues. Smaller SMEs below USD 10 million did not recognise Industry 4.0 benefits and were not on board due to various constraints with investment and operation scale as a hindrance. We observed that specific sectors such as hi-tech manufacturing, electronics, semiconductors, and medical equipment were very well positioned to benefit from Industry 4.0 initiatives and SMEs in these sectors.

"The writing is on the wall with COVID coming in. I think Industry 4.0 will get an impetus in terms of people wanting to adopt/ upgrade or maybe at least have the interest because traditional construction companies would not have changed their model."

(Interviewee 5)

5. Findings of SME Case Study Research: 'Racer Technology Pte Ltd'

Racer Technology Pte Ltd (founded in 1988) is a Singaporean medtech manufacturing and solutions firm, dedicated to the design of medical products, which produces devices and lab equipment.

The company later expanded into end-to-end solutions like surgical, patient care and diagnostic devices by harnessing their engineering design capabilities. With a big strength of loyal and dedicated staff being with the company for over 30 years (more than a quarter of them are enterprising research scientists and engineers), the company's founder and CEO, Mr. Koh, identified medtech as a growth sector and leveraged on the firm's strengths. From its humble origins as a contract manufacturer-driven medical technologies company, Racer Technology Pte Ltd has become a complete solutions provider that pursues a co-design collaborative approach with its partners.

5.1. *Being a Visionary*

Racer's conference rooms and the offices of the CEO are adorned with models of cars, vintage models of telephones, custom-designed items including card holders and wallets, which bear a resounding testament and significance to Mr. Koh's passion for engineering and his astute business acumen. The restlessness to constantly innovate, develop and/or improve on a product idea by revamping a design and creating an alternative approach are hallmarks of the markers laid down by Mr. Koh to ensure that his firm constantly collaborates and co-creates designs and solutions. This approach has consistently aided Racer to build partnerships globally with established firms and many of their leading partners, keep innovating and remain ahead of the curve. The company has been strongly riding on its engineering design capabilities to also cross-apply knowledge and expertise gained across various products. Racer's value propositions are high mix and low volume product options for customers, engineering design capabilities including co-design options, and finally being a reliable partner for leading global brands to develop their products with top quality and reliability.

5.2. *Overcoming Challenges*

Racer realised soon after its inception that a facility to prototype ideas would be critical while also understanding the longer gestation

periods in the industry with medical devices needing stringent testing and certification requirements. The company was ISO-14385
certified for quality compliance to allow it the freedom to design
and manufacture medical devices. This created another opportunity
for Racer to work with other local SMEs to manage certifications like
CE and regulatory approvals since Racer was very versatile with the
approach needed, including the complicated paperwork formalities
critical for such requirements.

To sharpen Racer's competitive edge, it was essential to develop a
strategy to delve into new product solutions by tapping into the available resources. Mr. Koh entered into a partnership with A*STAR's
research institute, Singapore Institute of Manufacturing Technology
(SIMTech), to research and build production solutions in partnership.

5.3. *Digital Factory*

During a business mission to Germany in 2016, Mr. Koh first-hand
witnessed Industry 4.0 initiatives, which were implemented by
German companies. He was so impressed that upon his return, he
stressed to his team on the benefits of getting Industry 4.0 established in Racer. It would have been only possible with a total
commitment from Mr. Koh's side to implement Industry 4.0 initiatives while taking his long-term employees into complete confidence
that this transformation would be beneficial to one and all:

> "I came back from the business mission to Germany and told my
> team, that I want Industry 4.0 implemented yesterday and how can
> we get this done. At that point in time, the Singapore market pre-
> dominantly comprised of overseas technology vendors of Industry
> 4.0 solutions, who were also very expensive. I wanted a local vendor
> who would understand my needs and be able to supply it while
> others can also use it at a later time by re-tailoring the solution to
> meet their customized requirement."

The identification of the right partner in 2016 to enter the
Industry 4.0 initiatives was another challenge since the technologies
were very new and Singapore was flooded with several new providers.

The background work to find the appropriate digital solution suitable for Racer was critical as not every Industry 4.0 solution was suitable or easily adaptable readily. Eventually, Racer tapped into its existing partnership with SIMTech on Industry 4.0, and since they already had some familiarity in understanding the process, it was easier for them to custom-design solutions, which could be thereafter easily integrated with Racer. As SIMTech was very well familiar with the grants and financial assistance schemes, it was very helpful for Racer to seek their assistance to understand areas where financial support was feasible due to technological upgradation. This resulted in a lot of relief in terms of the paperwork needed to be done in view of Racer's limited resources.

> "A lot of people mistake Industry 4.0 as automation. Automation is actually Industry 3.0. Industry 4.0 is essentially about the collection of the relevant data and information in a factory by using sensors, RFIDs etc. to make it a closed-loop system that allows us to have real-time information about our operations."
>
> — Mr Koh

Mr. Koh realised the urgent need for data-driven decisions to be able to improve their efficiency and productivity, and to keep labour costs down, while improving the output and reducing the unit costs of products to remain competitive. It has also aided in the improvement of the process by reducing the cycle time or frequency while optimising the setup time. With their A*STAR partnership, Racer could tap into their Model Factory initiative, which is designed to improve productivity by keeping costs lower through technological advancements in three major areas: operations management, equipment effectiveness and inspection.

The financial support provided by the government agencies like Enterprise Singapore allowed Racer to identify and gradually adopt four modular solutions specifically tailored to their needs devised by A*STAR. It helped Racer to improve operations efficiency by working with two planners against four planners and speeding up planning lead time from 2 weeks to 3 working days while doing away with the third shift completely. Overall, this enabled an efficiency enhancement of 70% with better machine utilisation which allowed flexibility to determine spare capacity easily due to the availability of data, when some orders could be slotted into the third shift on an urgent basis. The data collected from software greatly facilitate streamlining and improvement of operations.

On another front like inspection, the software aided the supervisors to verify any quality matters whether on-site or remote basis to facilitate immediate decision-making to be implemented, thus saving a lot of manual entry procedures as well as delayed decision-making. Overall, it is of great help to the implementation of the lean manufacturing approach, including real-time tracking of raw material shortages as well as managing Just-In-Time (JIT) supplies of inventory for Racer's manufacturing works.

With 1800 employees across eight factory locations, the digital factory approach has resulted in connected data advantages. All the eight factories at various locations are Industry 4.0-compliant and interconnected, thus allowing for flexibility across all locations to manufacture the various products. Singapore's excellent free trade arrangements and location allow Racer to strategically position itself while leveraging on their factories across Asia, as they continue to expand and foray into new markets.

5.4. *Talent Management*

Racer has many loyal staff who have spent over 30 years with the company. The all-important need to retain their rich experience while reskilling them to refresh their skill sets in line with Industry 4.0 initiatives was very well acknowledged by the leadership team to enable a successful transformation. Mr. Koh personally ensured that the right communication is sent to all the employees and announced that no retrenchments would be made while making full disclosure of the plans by the company to retrain and redeploy staff if the situation demanded. Mr. Koh values his employees as critical key resources to the extent that he claims to be expendable himself but not the long-term committed staff who manage his production lines and business, which shows the top management commitment and belief in the employees.

Around 15 staff were sent to the SIMTech facilities in 2017 to familiarize themselves with the modular solutions being implemented and benefit from an on-the-site training to observe the technological changes. Followed by this, two more staff underwent the Workforce Skills Qualification (WSQ) course to understand the microfluidics manufacturing process.

To fulfil its long-term needs for talent, Racer has initiated partnerships for internships with polytechnics in Singapore, the Institute of Technical Education (ITE) and also the Singapore University of Technology. The company offers internships to meet future manpower needs.

5.5. *Partnerships: A Collaborative Recipe for Success*

Right from Racer's inception, Mr. Koh firmly believed that relationships and partnerships were the only way for an aspiring SME to build and transform the business. It was evident in the form of associations formed with technology providers like SIMTech or Institutes of Higher Learnings (IHLs) for new recruits and internships, the partnerships with key customers to develop a test kit or even during the COVID-19 pandemic when the company was chosen as one of the first local SMEs to produce face shields and masks in Singapore by the government (it took the firm less than 1 week to deliver them). Later, the company was also involved in COVID-19 reagents and test kits (digital PCR kits) considering their immense experience in Ebola test kits in the past.

Racer believes firmly in the philosophy of developing products for established companies in Europe and the United States, which have a strong branding and a sound distribution network as an Original Brand Manufacturing (OBM) player. The need to market the ideas under its own brand name does not seem appealing to Mr. Koh as it is felt that Racer is only keen to leverage on its strength in design and ideas rather than marketing them. The long-term collaboration with international firms in the US and Japan also helps Racer to cross-pollinate knowledge and experience in one area to several other potential ideas, thus creating a potentially winning formula in view of the long gestation periods. This open-minded approach of sharing and trusted ecosystem partners allows Racer to serve as a mentor to guide start-ups and interested SMEs, thus creating valuable partnerships for both sides to cement their business.

Racer is now evaluating possibilities to collaborate with SIMTech to aim for waste reduction by developing 3D automated screening inspection stations. The open mindset to allow even industry participants to draw ideas and share success stories and elements from Racer's Industry 4.0 foray is a very refreshing outlook — this has been well documented in the German SME style of healthy collaboration even between competitors and is considered to be one of the ingredients to the success of Industry 4.0 in Germany.

5.6. *The Roadmap Ahead*

Racer has been one of the earliest adopters of Industry 4.0 initiatives in Singapore, which has allowed the firm to make rapid strides in smart manufacturing. In terms of revenue growth, the company is growing at an estimated rate of 11% year-on-year and is expecting to double revenues in the next 4–5 years. Research and Development (R&D) is a very key driver to the firm's success in remaining ahead of the competition curve. Racer has been consistently investing more than SGD 1 million with further increases expected in the years ahead to grow revenues and profits. Around six patents are consistently done on a yearly basis in the area of medical devices in line with this need to strive in the R&D space to grow the business.

The company is now actively engaging with SIMTech to also co-develop wearables like activity trackers. The race of innovation can be won with partners and collaborators according to Mr. Koh, and it seems that Racer is poised for a fascinating period ahead under his able leadership.

6. Conclusion

This chapter is part of a project that examines the readiness of Singapore-based SMEs in the manufacturing sector to implement

Industry 4.0 solutions. SMEs must adopt a clear strategy and define a roadmap to align their organisation and entire workforce with the goals of Industry 4.0. SMEs need to closely network with educational institutions, industry partners and the government (e.g., Singapore's A*STAR research institute SIMTech) to create strong ecosystem linkages for mutual benefits. While the Industry 4.0 transformation has significant top-down leadership support in Singapore, SME bosses and employees must buy-in into the challenging SME trans-formation process with clear support and actionable guidance to see Industry 4.0 initiatives succeed.

SMEs are increasingly getting convinced of the benefits of digi-talisation and automation. Digitalisation, analytics, and automation will be in significant demand, and firms providing the right work environment with a progressive mindset will reap the rewards in the times ahead. Singapore can play a critical role in ASEAN by leading

Table 2.   Impact of Industry 4.0 on Racer's Business Model Building Blocks

Business Model Components	Findings/Impact	Score
Value Proposition	Benefit(s) received by customers, measurable impact in creating customer value, software-centric business model.	
	Greater benefits achieved internally with regard to process, production, shipment, inspection and plan-ning resulting in the optimisation of labour and improved efficiency.	
Customer Segments	Identification of new customer segments and 'white space' opportunities in existing/new sectors.	
	Limited benefits as significant requirements of R&D, branding, exhibitions, and customer proximity remain key drivers.	
Channels	Identification of new digital channels for sales, distri-bution and communication.	
	Less accrual of benefits felt barring increased use of technology applications for paperless commercial transactions.	

(Continued)

Table 2. (*Continued*)

Business Model Components	Findings/Impact	Score
Customer Relationships	Collaborative partnerships including R&D initiatives to create innovative ideas, platform models, co-creation, etc.	
	Good collaborative exercises with partners and proximity to customers has resulted in joint lab experiments to develop new product ideas.	
Key Activities	Approach toward new R&D, product and process design, servitisation with new systems and metrics to be more effective.	
	Significant adoption of new software and technology initiatives resulted in landmark improvements in production and process efficacy.	
Key Resources	Human capital alignment, qualifications requirement and tackling mindset issues.	
	Significant challenges overcome by excellent handling of employees combined with training to adopt new initiatives proactively.	
Key Partners	Collaborative ecosystem of partners including technology providers, consultants, vendors, etc. and outcomes.	
	Excellent leverage of the ecosystem with partners like A*STAR and SIMTech at an early stage resulted in close partnerships to co-develop solutions.	
Revenue Streams	Introduction of new models for revenue generation like pay-by-usage models (SaaS, DaaS), etc. and improvements in revenue generation streams.	
	Less benefits accrued thus far, but efforts are ongoing.	
Cost Structure	Capital investment, software costs, etc. have affected production-related costs and improved cost structure.	
	While IT costs shot up, production costs including operator monitoring costs became more efficient with better real-time cost structure accuracy.	

Scale:

0	20	40	60	80	100
No impact	Minimal impact	Fair impact	Good impact	Significant impact	Maximum impact

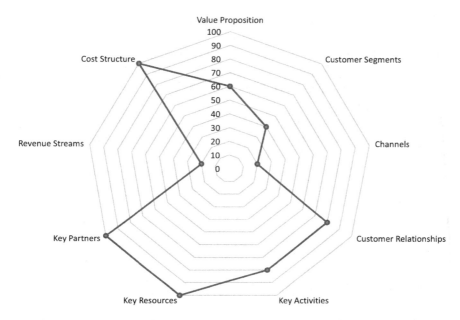

Figure 2. Impact of Industry 4.0 on Racer's Business Model Building Blocks

Industry 4.0 initiatives such as advanced manufacturing, innovation, and digital activities across the entire value chain. As Table 2 and Figure 2 suggest, Industry 4.0 can have a positive impact on the business models of local SMEs, provided it is executed by a proactive SME owner-manager (Bateman & Grant, 1993) such as Racer Technology's 'boss' who is able to create and capture new value in the form of data-driven process, production, shipment, inspection and planning improvements based on a compelling value proposition and collaborative value network.

References

Bateman, T. S. & Crant, J. M. (1993). The proactive component of organizational behavior: A measure and correlates. *Journal of Organizational Behavior*, 14(2), 103–118. DOI: 10.1002/job.4030140202.

Belton, K. B. *et al.* (2019). Who will set the rules for smart factories? *Issues in Science & Technology*, 35(3), 70–76.

Eisenhardt, K. M. & Martin, J. A. (2000). Dynamic capabilities: What are they? *Strategic Management Journal*, 21(10–11), 1105–1121. https://doi.org/10.1002/1097-0266(200010/11)21:10/11<1105::AID-SMJ133>3.0.CO;2-E.

Eisenhardt, K. M. & Graebner, M. E. (2007). Theory building from cases: Opportunities and challenges. *Academy of Management Journal*, 50(1), 25–32.

Gioia, D. A. *et al.* (2013). Seeking qualitative rigor in inductive research: Notes on the gioia methodology. *Organizational Research Methods*, 16(1), 15–31. DOI: 10.1177/1094428112452151.

Girotra, K. & Netessine, S. (2014). Four paths to business model innovation. *Harvard Business Review* (July–August 2014) 1 July. Available at: https://hbr.org/2014/07/four-paths-to-business-model-innovation (Accessed: 29 September 2020).

Kagermann, H., Wahlster, W. & Helbig, J. (2013). Recommendations for Implementing the Strategic Initiative Industrie 4.0 — Final Report of the Industrie 4.0 Working Group. Communication Promoters Group of the Industry-Science Research Alliance, acatech, Frankfurt am Main. Available at: https://www.din.de/blob/76902/e8cac883f42bf28536e7e8165993f1fd/recommendations-for-implementing-industry-4-0-data.pdf.

Launch of A*STAR's Model Factory @ ARTC (nd). Available at: https://www.a-star.edu.sg/News-and-Events/a-star-news/news/publicity-highlights/launch-of-a-star-s-model-factory—artc (Accessed: 26 September 2020).

Marr, B. (2016). What Everyone Must Know About Industry 4.0. Available at: https://www.forbes.com/sites/bernardmarr/2016/06/20/what-everyone-must-know-about-industry-4-0/#66508a40795f (Accessed: 26 September 2020).

McKewen, E. (nd). What is Smart Manufacturing? (Part 1A). Available at: https://www.cmtc.com/blog/what-is-smart-manufacturing-part-1a-of-6 (Accessed: 26 September 2020).

Miles, M. B. & Huberman, A. M. (1984). Drawing valid meaning from qualitative data: Toward a shared craft. *Educational Researcher*, 13(5), 20–30. https://doi.org/10.3102/0013189X013005020.

Müller, J. M. (2019). Assessing the barriers to Industry 4.0 implementation from a worker's perspective. *IFAC-PapersOnLine*, 52(13), 2189–2194. DOI: doi.org/10.1016/j.ifacol.2019.11.530.

Osterwalder, A., Pigneur, Y. & Clark, T. (2010). *Business Model Generation: A Handbook for Visionaries, Game Changers, and Challengers.* Strategyzer series. Hoboken, NJ: John Wiley & Sons.

Schwab, K. (2016). The Fourth Industrial Revolution: What it Means and How to respond. Available at: https://www.weforum.org/agenda/2016/01/the-fourth-industrial-revolution-what-it-means-and-how-to-respond/ (Accessed: 26 September 2020).

Singapore's Companies take Leap into Advanced Manufacturing (nd). Available at: https://www.edb.gov.sg/en/news-and-events/insights/manufacturing/factory-forward-advanced-manufacturing-takes-root-in-singapore.html (Accessed: 26 September 2020).

Taran, Y., Goduscheit, R. C. & Boer, H. (2019). Business model innovation — A gamble or a manageable process? *Journal of Business Models,* 7(5), 90–107.

Teece, D. J. (2010). Technological innovation and the theory of the firm: The role of enterprise-level knowledge, complementarities, and (dynamic) capabilities. In: *Handbook of the Economics of Innovation* (Vol. 1, Chapter 16) edited by B. H. Hall and N. Rosenberg. Amsterdam: Elsevier.

Zezulka, F. *et al.* (2016). Industry 4.0 — An introduction in the phenomenon. *IFAC-PapersOnLine,* 49(25), 8–12. DOI: https://doi.org/10.1016/j.ifacol.2016.12.002.

Chapter 4

AI Competency Acquisition Online: Engaging Undergraduate Students in an AI 101 Course through a Chatbot Workshop

Thomas Menkhoff and Lydia Ying Qian Teo

1. Introduction

In recent years, digital transformation has dominated industries at an unprecedented rate. Alongside the proliferation of Artificial Intelligence (AI) technologies in the workplace, institutions of higher learning are experiencing an unprecedented push to integrate AI into the education ecosystem (Popenici & Kerr, 2017; Renz & Hilbig, 2020). AI in education (AIED) has the propensity to enrich teaching and learning in higher education by personalising students' learning courses, automating assessment tasks, or providing 24/7 access to learning resources (Karandish, 2021). According to estimates by the AI Market in the US Education Report, the AIED market will grow at a CAGR of 47.77% during the period 2018–2022 (Report, 2018).

A popular application of AIED is deploying chatbots as conversational agents to engage students by supporting their learning beyond the classroom and enhancing their learning process. For instance, chatbot prototypes can be integrated with e-learning platforms, using NLP to interact with students by interpreting their queries and drawing up relevant information from the knowledge base module to assist students in their revision (Clarizia *et al.*, 2018).

Chatbots are also a viable solution to bridge the communication gap between students and instructors for distance learning (Tamayo *et al.*, 2020) and are effective when deployed appropriately by educators in areas such as teaching and learning foreign languages (Kim *et al.*, 2019; Nghi *et al.*, 2019). They are also suitable pedagogical tools to equip students with the necessary knowledge and skills to better understand emerging AI technologies and how they are applied beyond the classroom.

As educators, we are cognisant of the importance of the effects of motivation and student engagement on learning outcomes. Navarro *et al.* (2020) liken motivation to a resource that can be tapped into as a 'natural source of learning' to spur the achievement of learning objectives. Similarly, Monteiro *et al.* (2015) consider motivation to be the crucial factor in terms of performance and engagement of students.

In this paper, we report on experiences during the implementation of an experiential chatbot workshop integrated into an introductory undergraduate management course on 'Doing Business with AI.' at Singapore Management University (SMU) that provides non-IT students with an opportunity to build a chatbot prototype using the 'Dialogflow' program. One basic premise is the assumption that such a novel, learning outcome-related, hands-on chatbot workshop motivates and engages learners provided its pedagogical approach is effective.

The research questions for this study are as follows:

RQ1: To what extent does the experiential chatbot workshop result in student motivation and engagement?

RQ2: To what extent do motivation and engagement enable the achievement of desired learning outcomes for the workshop?

Besides sharing how the chatbot workshop was designed and implemented, we share initial qualitative feedback obtained from students via end-of-course evaluations and ongoing semi-structured interviews. These preliminary results suggest that experiential learning by doing indeed has a positive effect on students' learning. We also present some hypotheses from our ongoing work which will be addressed subsequently in a follow-up survey.

2. Dimensions of AIED

2.1. *Experiential Learning*

Experiential learning refers to a process that is designed for learners to acquire knowledge and skills by 'doing.' Having critically reviewed the teaching experience from past terms, the instructors of this module found that incorporating 'hands-on' elements and activities where students can apply their learning forms a better learning experience for students.

In view of the nascent space of AIED and technology-mediated learning (TML) at the course level, the design of the experiential chatbot workshop was influenced by learning theories such as Kolb's Experiential Learning Model (Sims, 1983) and Gagne's Nine Events of Learning (Northern Illinois University, n.d.). One key goal was to create a novel and interesting learning experience for students to foster better student engagement by deploying an engaging, activity-based instructional method.

One learning objective of MGMT240 pertains to enabling students to understand and explain the workings and applications of AI in various business functions:

> "Explain how private and public sector organizations engaged in customer service management, finance, marketing, supply chain management and manufacturing use machine learning, deep learning, neural networks, image analysis etc. to potentially become more competitive and 'effective' in real- time."

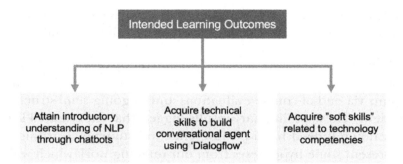

Figure 1. Intended Learning Outcomes of Experiential Chatbot Workshop

The intended learning outcomes of the experiential chatbot workshop can be categorised into three main pillars of competencies as illustrated in Figure 1.

To guide our teaching efforts during the workshop and to assess students' chatbot-related learning, we utilised Bloom's Taxonomy as explained in Table 1. Table 2 presents the learning outcomes.

One key learning outcome of the workshop is to enable students to demonstrate a basic level of understanding of Natural Language Processing (NLP), the technology that facilitates interaction between natural human language and computers, as exemplified by conversational agents like chatbots. Aligned with the popularisation of chatbot usage beyond the classroom (Deloitte Digital, 2018; Han, 2021), skills related to creating chatbots, or even understanding how chatbots are developed, are deemed as competencies valued in the workplace (Brynjolfsson & McAfee, 2017; Davenport & Ronanki, 2018; Zawacki-Richter *et al.*, 2019). Hence, the experiential workshop provides students with an opportunity to acquire skills to build a chatbot prototype using the 'Dialogflow' program during the hands-on activity. Experience using the Dialogflow will form the basis of knowledge and familiarity, making it easier for students to pick up or utilise other chatbot platforms or programs in the future. The workshop and the experiential learning activity were designed to impart students with other relevant skills such as conversation design and user-centric design that are transferrable to other situational contexts beyond the course.

Table 1. Bloom's Taxonomy of Learning Objectives Applied to our Experiential Chatbot Workshop

Bloom's Learning Objectives	Examples of Learning Objectives
Create	• Reproducing conversational agents on *Dialogflow* independently. • Generating chatbot prototypes using skills acquired from chatbot workshop for other contexts (organisational or project settings) using other mediums or platforms.
Evaluate	• Explaining what differentiates a good chatbot from an average chatbot, with some elaboration to substantiate the idea by giving relevant examples or providing detailed descriptions. • Assessing important considerations for implementing chatbots in various contexts (e.g., budget, maintenance requirement, and area of deployment). • Critiquing one's own chatbot to identify areas for improvement when reviewing and testing the prototype. • Identifying strengths and weaknesses in peer-presented work and providing constructive feedback for chatbot tested by volunteers.
Analyse	• Differentiating various approaches for the creation of chatbot prototypes and implementation in different organisational settings based on various considerations, such as the needs of the organisation(s), purpose of the chatbot, and appropriate chatbot personality.
Apply	• Enacting needs assessment to identify end-user requirements and scope of chatbot functions in the context presented (freshman orientation). • Using the logical diagram to map out chatbot conversation flow. • Implementing *Dialogflow* functions ('intent,' 'training phrases,' 'entities,' 'response,' and 'fallback') independently during the hands-on activity.
Understand	• Explaining the relationship and connection between AI, machine learning, and NLP, and discussing how NLP may have been applied in *Dialogflow* for speech recognition, natural language understanding, and eventually natural language generation (in voice or text output).

(*Continued*)

Table 1. (*Continued*)

Bloom's Learning Objectives	Examples of Learning Objectives
	• Learning that *Dialogflow* is a platform that can be used to create and test the prototype of conversational agents, and there are similar products and mediums available in the market which do not require in-depth programming skills to build chatbots. • Making sense of the logical flow diagrams and how they can be adapted to design conversation flow as part of the chatbot building process. • Interpreting the use of end-user needs assessment and importance of user-centric design in the creation of AI technology solution.
Remember	• Defining what a chatbot is. • Describing different types of chatbots available on the market. • Recognising examples of chatbots encountered in day-to-day life. • Connecting knowledge on chatbots and types of chatbots to real-life examples and use cases (e.g., which type of chatbot is more suitable for a specific type of organisation).

Table 2. Lesson Plan for Experiential Chatbot Workshop

Part I: Introduction (Warm-Up)		Duration: 30–45 min
1. Introduction on chatbots 2. Types of chatbots 3. Introduction to Dialogflow, NLP 4. Dialogflow demonstration (using a first-year student orientation case study)	**Content covered:** • Types of chatbots • How to select chatbots • Technology involved (NLP) • Dialogflow functions • Conversation design • User requirements, user-centric design, needs assessment	**Teaching Objectives:** – Overview on chatbot options and technology powering its capabilities. – Demonstrate real-life application of chatbots using case studies.

Table 2. (*Continued*)

Part II: Class Activities		Duration: 45 min
Activity 1 (25 min)	**Content covered:**	**Teaching Objectives:**
1. Create conversational agent on Dialogflow.	• Design of conversation flow on chatbot.	– 'Hands-on' segment on Dialogflow to practice conversation design.
2. Test chatbot prototype	• Step-by-step process of creating chatbot on Dialogflow (coding not required for workshop)	– Create a complete set of ('Intents,' 'Entities,' 'Response,' and 'Fallback')
3. Volunteers to present		– Students to test their chatbot agent
4. In-class discussion	• User-centric design	
Activity 2 (20 min)		– Critical thinking: application of chatbots in different contexts (beyond example given in class)
Discussion: Implementing chatbot in the workplace		

2.2. *Motivation*

Motivation has been identified as a crucial factor central to student engagement and performance (Monteiro *et al.*, 2015). Hence, it is examined as a key variable in this research study. We consider motivation at the process stage of Winkler and Söllner's input–process–output model (2018) as one's motivation will influence and affect the learning experience during a practical workshop designed to develop a particular AI-related skill such as NLP.

A well-researched theoretical framework for studying motivation is the Self-Determination Theory (SDT) by Ryan & Deci (2000). Intrinsic motivation refers to doing things 'for their own sake' or acting as if the task is perceived to be inherently interesting or pleasant to the individual. Intrinsic motivation is associated with positive learning outcomes in formal education such as school performance and achievements (Augustyniak *et al.*, 2016; Heindl, 2020).

The SDT posits that humans have proactive tendencies that are manifested in 'learning, mastery and connection with others' (Ryan & Deci, 2020). In an educational setting, basic psychological needs for autonomy, competence, and relatedness need to be met for effective learning to occur. In our chatbot workshop, we tried to experiment with creating such a needs-supportive setting to harness this resource.

In designing our data collection tools, we used the Intrinsic Motivation Inventory (IMI) as it is derived from the Self-Determination Theory. As a valid and reliable measurement instrument, it has been widely used to measure intrinsic motivation (Leng *et al.*, 2010; Monteiro *et al.*, 2015; Augustyniak *et al.*, 2016; Heindl, 2020; Navarro *et al.*, 2020).

The IMI was deemed as a suitable scale to be incorporated into our survey instrument as its items consists of simple, short sentences which are easily comprehensible for respondents (tertiary students). While other studies have found IMI unsuitable and ineffective as the items may be too complex for respondents, such as elementary school students who are young with lower levels of literacy (Leng *et al.*, 2010), we did not come across any such issues. While some items were modified to present the context of the experiential chatbot workshop, we used five of the most relevant subscales of intrinsic motivation in our survey questionnaire. They consist of *interest/enjoyment, perceived competence, effort/importance, pressure/tension, and value/usefulness.* The subscales of *perceived choice* and *relatedness* were excluded (Ryan & Deci, 2020; Navarro *et al.*, 2020). This was done because the experiential chatbot was a compulsory component of the course syllabus, which all students had to attend (i.e., they had no *choice*). Similarly, *relatedness* was perceived to have a minimal role as the students mostly interacted with the workshop instructor over a short duration only.

2.3. *Engagement*

Student engagement is closely linked to learning outcomes and, according to Mandernach (2015), is an 'integral component' for learning effectiveness. Other research in this domain established that higher levels of engagement often translate into better learning outcomes, as engaged students are 'good learners' (Handelsman *et al.*, 2005). Existing literature studies engagement across different educational levels and indicates that the learner's level of engagement can vary due to intrinsic factors, or 'learner variables,' such as individual motivation, and other extrinsic factors that influence the learning process, such as the involvement and quality of instructors (Handelsman *et al.*, 2005; Jung & Lee, 2018).

During the experiential chatbot workshop, we tried to foster a safe and open learning environment for collaborative learning. Learners could ask questions and interact with the instructor or peers to respond to any topic-related questions or comments at any point during the workshop. Similarly, students were encouraged to volunteer and present their work to the class with others invited to share constructive feedback in the spirit of collaborative learning. The workshop also included attention checkers for students to indicate their stage of progress during the guided demonstration, and they could make use of both the audio/video and text (chat box) functions to share their input or reach out to the instructor for help.

Engagement is a multidimensional construct (Handelsman *et al.*, 2005; Mandernach, 2015). To effectively examine and measure student engagement, it is important to choose an instrument that is suitable for the scope of our 'unique context' (Mandernach, 2015) in this research. We selected the Student Course Engagement Questionnaire (SCEQ) as the scale of reference because the 23-item measure is easy to administer. Items from the SCEQ were modified slightly to fit the context of the workshop. More importantly, the SCEQ is robust and the items capture four dimensions of engagement which includes (i) skills engagement, (ii) participation/interaction, (iii) emotional engagement, and (iv) performance engagement (Handelsman *et al.*, 2005; Nasir *et al.*, 2020), giving instructors deeper insights into student engagement beyond the information that can be observed from student's in-class responses during the workshop and offering hindsight assessments and inferences based on their grades (Handelsman *et al.*, 2005).

Figure 2 illustrates our conceptual model which we intend to corroborate during the quantitative stage of our ongoing research study. Its development is based on the learning theories and TML models introduced earlier aimed at measuring the effects of the hands-on chatbot workshop on students' engagement and

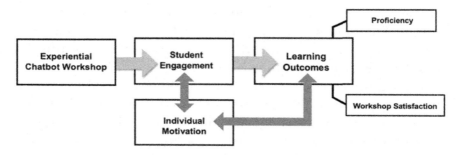

Figure 2.   Conceptual Model for Engagement, Motivation, and Learning Outcomes

motivation as drivers of acquiring AI-related competencies such as the understanding and application of NLP.

The following hypotheses were derived from the conceptual model illustrated in Figure 2:

> **H1**: Students with a high level of individual motivation will be more engaged as a result of the Chatbot workshop than students who have a low level of individual motivation.

> **H2**: Students with a high level of engagement as result of the chatbot workshop experience will be more motivated than students who have a low level of engagement.

> **H3**: Students with a low level of pre-workshop (functionality-related) proficiency will be more motivated as result of the chatbot workshop than students with a high level of pre-workshop (functionality-related) proficiency.

> **H4**: Students with a high level of engagement as result of the chatbot workshop experience will have a higher level of chatbot-related competencies than students who have a low level of engagement.

> **H5**: Students with a high level of engagement as result of the chatbot workshop experience will be more satisfied with the workshop than students who have a low level of engagement.

> **H6**: Students with a high level of individual motivation will be more satisfied with the Chatbot workshop than students who have a low level of individual motivation.

H7: Students with a high level of individual motivation will report a higher level of chatbot-related competencies than students who have a low level of individual motivation.

H8: Students with a high level of pre-workshop (functionality-related) proficiency will have a higher level of post-workshop competencies than students with a low level of pre-workshop (functionality-related) proficiency.

H9: Students with a low level of pre-workshop (functionality-related) proficiency will be more engaged because of the chatbot workshop than students with a high level of pre-workshop (functionality-related) proficiency.

H10: Students with a low level of pre-workshop (functionality-related) proficiency will be more satisfied with the chatbot workshop than students with a high level of pre-workshop (functionality-related) proficiency.

H11: Students with a high level of alignment will be more satisfied with the Chatbot workshop than students who have a low level of alignment.

3. Methods

The sort of 'intervention' in our research is a 90-minute experiential chatbot workshop which was conducted online during a regular class session in Term 1 of Academic Year 2021–22. Due to the university's COVID-19 regulations, chatbot-mediated learning was facilitated during a synchronous class session via Zoom by using a structured set of teaching and learning materials (Winkler & Söllner, 2018). The workshop activities were not graded to encourage inquiry-based learning and active participation. Removing any graded components was motivated by the instructor's desire to exclude extrinsic motivation factors (Leng *et al.*, 2010; Zaccone & Pedrini, 2019; Deci & Ryan, 2020; Navarro *et al.*, 2020) that may exert different pressures on the learning process such as participating only for grades.

Table 3.   Student Demographics

N	Gender	Year in School	Affiliated School	Major
43	23 Females	4 Year 1	0 Accountancy	8 Business Management
	20 Males	13 Year 2	43 LKCSB	13 Finance
		14 Year 3	0 Computing & IS	8 Marketing
		11 Year 4	0 Economics	5 Ops. Management
		1 Year 5	0 Law	4 Strategic Management
			0 Social Sciences	5 Others
	43	43	43	43

An online post-workshop survey was administered to all students enrolled in the course 'MGMT240 Doing Business with AI' Table 3 provides details of the study population.

Students enrolled in this course are of different course levels (year of study) and pursue different majors as indicated in Table 3.

Survey questions focus on potential workshop aspects students might have enjoyed with a view towards highlighting areas for improvement and enhancing their learning effectiveness. The responses to several open-ended questions will be studied thoroughly and incorporated into future, further improved runs of the workshop.

The questionnaire was accessible online over a 7-day period, and students were encouraged to participate on a voluntary basis. All collated responses from the students were anonymous, and respondents were assured that the survey data would in no way affect their course grades. Altogether, 43 students completed the online survey (23 female, 20 male) with a response rate of 97.7%. The majority of the surveyed chatbot workshop participants (81.4%, n=35) reported to be engaged of which 18.6% (n=8) were highly engaged. The analysis of this dataset has yet to be conducted. At the moment, we are conducting several half-structured interviews with student volunteers to explore their experiences and to assess the overall impact of the workshop. The main study variables have been introduced earlier (please refer to Figure 2).

A quick check of the data indicates that students enjoyed acquiring new AI-related skill(s) through the hands-on experiential learning activity. They also rated the instructor and the instructional method as positive.

4. Results and Discussion

Our interim study results suggest that our hands-on experiential chatbot workshop design managed to engage and motivate students to acquire new AI-related skills such as NLP as indicated by the following comments from students:

"It gave us a first-hand experience on NLP. We kept talking about NLP in class, but we do not know exactly what it is or how it looks like, how it is being used and applied."

"Before this module, I had low to almost zero knowledge of Artificial Intelligence, and I am glad I took this module as I have learnt a lot about AI including ML, DL, NLP, Chatbots, etc. ... Also, not only do I have a better understanding of the different types of AI technologies, I also understand the history and the impact of AI. ... Overall, I enjoyed the module, and this module definitely did spark my interest in AI, definitely a suitable AI 101 module for students."

While not representative, these comments corroborate the logic of our conceptual model as illustrated in Figure 3.

An interesting aspect of the qualitative feedback relates to the pressure/tension dimension that affects intrinsic motivation, i.e., the perception that the learning experience during the chatbot workshop was generally positive '*with little to no pressure to perform or excel.*'

The general sentiment gathered from the semi-structured student interview data suggests that participants found the live demonstration and hands-on activity to be highly engaging as students got to interact and learn from both the instructor and their peers. Engagement from the workshop seems to translate into positive learning outcomes for the participating students as the following comments suggest:

"... the workshop has been very engaging ... my mind was constantly working, and engagement was very high even if I might not have asked as many questions as I might have liked."

"The engagement has made things very clear as people got to make suggestions, and we had to go over it in class, even questions I haven't thought of, we would go over. Even now, I remember the five steps that were taught during the workshop. This means that it was very useful in promoting memory retention, I would say…"

From the interviews, we also observed that students with varying levels of motivation prior to attending the workshop reflected on having experienced motivation during the workshop due to certain attributes that range from useful concepts such as the mind map for user needs assessment and practical workshop approach to the course design and lesson materials.

"My interest towards the subject is already very high, but my motivation was a derivative of what I saw (authors: during the workshop). I saw many demo questions of the new SMU student on the slides … looking at that I immediately thought 'how can I apply that to things that I probably want to create on my own' …, 'what questions I would ask' and that could be a motivation (for me) to make sure that I got everything down."

"Before the workshop, I wasn't very motivated as I wasn't aware of the technology behind what a chatbot is and the benefits and uses of the workshop. During the workshop, I became more motivated because I was able to understand more about how a chatbot works and why people use chatbots…my motivation increased overall."

Students we talked to found the workshop to be aligned with their learning expectations and learning goals. Beyond the scope of the workshop, students felt that the MGMT240 module had some positive impact on their learning in general.

Table 4. Bloom's Taxonomy of Learning Outcomes Applied to our Experiential Chatbot Workshop

Bloom's Learning Objectives	Examples of Learning Outcomes
Create	• Create new agents into the Dialogflow web console. • Build working chatbot prototype independently (no coding required).
Evaluate	• Assess how responses could be tailored for specific contexts based on user requirements and to communicate effectively with end-users. • Differentiate between the adoption of different chatbot personalities for different contexts (e.g., customer service chatbot versus corporate enquiries). • Compare the effectiveness and viability of the logical flow diagram and conversation design against those prepared by peers.
Analyse	• Organise the logical flow and conversation flow of the chatbot prototype into a draft and transfer the design into *Dialogflow*. • Identify areas where the logical flow and conversation flow may be incoherent or ineffective and enact necessary adjustments to refine the performance of the chatbot prototype.
Apply	• Draw a simple representation of the conversational chatbot workflow. • For the context presented, implement *Dialogflow* functions taught during the workshop: (i) craft intents, (ii) create a list of training phrases for the corresponding intents, (iii) create suitable entities, (iv) prepare corresponding responses, and (iv) create suitable fallback intents on the *Dialogflow* web console.
Understand	• Distinguish between menu-button-based, keyword recognition-based, and smart AI-enabled chatbots that can simulate near-human interactions with customers. • Explain why the 'right' choice of chatbots depends on the target audience and the help they are hoping to get from the chatbot. • Explain why an effective chatbot requires NLP and other deep learning techniques to understand the underlying intent of human language.

(*Continued*)

Table 4.   (*Continued*)

Bloom's Learning Objectives	Examples of Learning Outcomes
	• Explain the function of the *Dialogflow* (virtual) agent, who handles all user conversation for a given use case, translating voice or text-based input into structured data that your application can understand and analyse to act on the user requests and/or fetch data to answer questions.
	• Explain that *Dialogflow* intent categorises the purpose of one side of a conversation turn (usually the user) so the other side (usually the agent) can act and respond accordingly.
	• Understand that for each intent, many training phrases (variations of different users may express the same idea, e.g., make the same enquiry) need to be created on *Dialogflow* to teach the conversational agent to understand and recognise the intent such that it can produce a relevant outcome or action.
	• Recognise that entities are specific parts of user input that describe some useful information (parameters) that *Dialogflow* can extract and match to perform a particular action.
	• Explain the importance of providing response information on *Dialogflow* for each intent when building a conversational agent.
	• Recognise that the purpose of fallback intents on Dialogflow is to help the conversation continue because the machine itself cannot recognise user input (intents).
Remember	• Identify different types of chatbots and their uses.
	• Point out that Dialogflow is an NLP program that can be used to build conversational applications for different end-users.
	• Recognise that NLP enables smart chatbots to read text, hear speech, interpret it, measure sentiment, and determine which parts are important.

"My learning effectiveness increased … This live workshop helps me to improve my understanding of concepts I have learned in class, and it is good in terms of the syllabus and doing well in terms of the course in general."

Our overall workshop experience suggests that successful student engagement in a digital AI context benefits from the embeddedness in a trusted T&L community comprising both (learning) instructors and learners energised by a robust knowledge-sharing culture.

5. Conclusion

This paper presents case study-based insights into the learning objectives and expected learning outcomes of a hands-on chatbot workshop designed as an engaging teaching tool in an introductory AI course taught by the authors at Singapore Management University (SMU). We introduced the pedagogy and instructional strategies of the experiential chatbot workshop and shared how they were derived from relevant learning theories to ensure the acquisition of respective competencies and to help learners understand why chatbot-related knowledge and skills are useful to them. We developed and corroborated a conceptual model indicating how the hands-on chatbot workshop might lead to outcomes such as motivation and engagement, eventually resulting in the achievement of the desired learning outcomes. From the interim findings, we can deduce that intrinsic motivation is an important factor that affects learning effectiveness. We expect that students with a high level of individual motivation will be more engaged as result of the chatbot workshop than students who have a low level of individual motivation (H1). A full analysis of our hypotheses has yet to be conducted.

One limitation of our ongoing study is the small sample size ($N = 43$), with 100% of the respondents enrolled in the same school (Lee Kong Chian School of Business). Even though respondents belong to different course levels and majors, it is possible that the sample variation may be limited and insufficient for researchers to understand the implications at cross-disciplinary levels. As all respondents were students enrolled in one course, there might also be a sample selection bias in the study. The lack of a control group limits the generalisation of the findings — something which could be addressed in future research based on a larger sample size.

Another viable approach to expand the scope of this study would be to measure dimensions of intrinsic motivation such as *perceived choice* and *relatedness* and explore to what extent these factors influence learning effectiveness and affect learning outcomes. In a follow-up study, researchers could also evaluate possible effectiveness-related differences between an experiential chatbot workshop that is conducted face-to-face (F2F) and one that is facilitated online via Zoom.

Acknowledgement

The authors gratefully acknowledge that this work was supported by a Ministry of Education (MOE) Tier 1 grant (MSS20B015).

References

Augustyniak, R. A., Ables, A. Z., Guilford, P., Lujan, H. L., Cortright, R. N. & DiCarlo, S. E. (2016). Intrinsic motivation: An overlooked component for student success. *Advances in Physiology Education*, 40(4), 465–466. https://doi.org/10.1152/advan.00072.2016.

Brynjolfsson, E. & McAfee, A. (2017). What's driving the machine learning explosion? Three factors make this AI's moment. Harvard Business Review. https://hbr.org/2017/07/whats-driving-the-machine-learning-explosion.

Clarizia, F., Colace, F., Lombardi, M., Pascale, F. & Santaniello, D. (2018). Chatbot: An education Support system for student. *Cyberspace Safety and Security*, 11161, 291–302. https://doi.org/10.1007/978-3-030-01689-0-23.

Davenport, T. H. & Ronanki, R. (2018). Technology — Artificial intelligence for the real world. Harvard Business Review, January–February Issue. https://hbr.org/2018/01/artificial-intelligence-for-the-real-world.

Deci, E. L. & Ryan, R. M. (2008). Self-determination theory: A macrotheory of human motivation, development, and health. *Canadian Psychology*, 49(3), 182–185.

Han, M. C. (2021). The impact of anthropomorphism on consumers' purchase decision in Chatbot Commerce. *Journal of Internet Commerce*, 20(1), 46–65. https://doi.org/10.1080/15332861.2020.1863022.

Handelsman, M. M., Briggs, W. L., Sullivan, N. & Towler, A. (2005). A measure of college student course engagement. *The Journal of Educational Research*, 98(3), 184–192. https://doi.org/10.3200/JOER.98.3.184-192.

Heindl, M. (2020). An extended short scale for measuring intrinsic motivation when engaged in inquiry-based learning. *Journal of Pedagogical Research*, 4(1), 22–30. https://doi.org/10.33902/JPR.2020057989.

Jung, Y. & Lee, J. (2018). Learning engagement and persistence in massive open online courses (MOOCS). *Computers and Education*, 122, 9–22. https://doi.org/10.1016/j.compedu.2018.02.013.

Karandish, D. (2021). 7 benefits of AI in education. The Journal, Transforming Education through Technolgy, https://thejournal.com/articles/2021/06/23/7-benefits-of-ai-in-education.aspx (Accessed on 7 November 2021).

Kim, N., Cha, Y. & Kim, H. (2019). Future English learning: Chatbots and artificial intelligence. *Multimedia-Assisted Language Learning*, 22(3), 32–53. https://search.library.smu.edu.sg/permalink/65SMU_INST/1ba19kd/cdi_kiss_primary_3703643.

Leng, E. Y., Wan Ali, W. Z. B., Baki, R. & Mahmud, R. (2010). Stability of the Intrinsic Motivation Inventory (IMI) for the use of Malaysian form one students in ICT literacy class. *Eurasia Journal of Mathematics, Science and Technology Education*, 6(3), 215-226. https://doi.org/10.12973/ejmste/75241.

Mandernach, B. J. (2015). Assessment of student engagement in higher education: A synthesis of literature and assessment tools. *International Journal of Learning, Teaching and Educational Research*, 12(2), 1–14.

Monteiro, Mata, L. & Peixoto, F. (2015). Intrinsic motivation inventory: Psychometric properties in the context of first language and mathematics learning. *Psicologia, Reflexão e Crítica*, 28(3), 434–443. https://doi.org/10.1590/1678-7153.201528302.

Nasir, M. A. M., Janikowski, T., Guyker, W. & Wang, C. C. (2020). Modifying the student course engagement questionnaire for use with online courses. *The Journal of Educators Online*, 17(1).

Navarro, O., Sanchez-Verdejo, F. J., Anguita, J. M. & Gonzalez, A. L. (2020). Motivation of university students towards the use of information and communication technologies and their relation to learning styles.

International Journal of Emerging Technologies in Learning, 15(15), 202–218. https://doi.org/10.3991/ijet.v15i15.14347.

Nghi, T., Huu Phuc, T. & Nguyen Tat, T. (2019). Applying AI Chatbot for teaching a foreign language: An empirical research. *International Journal of Scientific & Technology Research*, 8. 897–902.

Northern Illinois University. (n.d.). Gagné's nine events of instructions. Centre for Innovative Teaching and Learning. https://www.niu.edu/citl/resources/guides/instructional-guide/gagnes-nine-events-of-instruction.shtml

Popenici, S. & Kerr, S. (2017). Exploring the impact of artificial intelligence on teaching and learning in higher education. *Research and Practice in Technology Enhanced Learning*, 12(1). https://doi.org/10.1186/s41039-017-0062-8.

Renz, A. & Hilbig, R. (2020). Prerequisites for artificial intelligence in further education. *International Journal of Educational Technology in Higher Education*, 17(1), 1–21. https://doi.org/10.1186/s41239-020-00193-3.

Report. (2018). Artificial intelligence market in the US education sector 2018–2022. https://www.researchandmarkets.com/reports/4613290/artificial-intelligence-market-in-the-us?utm_source=BW&utm_medium=PressRelease&utm_code=5lshzz (Accessed on 7 November 2021).

Ryan, R. M. & Deci, E. (2000). Self-determination theory and the facilitation of intrinsic motivation, social development, and well-being. *The American Psychologist*, 55, 68–78. 10.1037/0003-066X.55.1.68.

Ryan, R. M. & Deci, E. L. (2020). Intrinsic and extrinsic motivation from a self-determination theory perspective: Definitions, theory, practices, and future directions. *Contemporary Educational Psychology*, 61, 101860. https://doi.org/10.1016/j.cedpsych.2020.101860.

Sims, R. (1983). Kolb's experiential learning theory: A framework for assessing person-job interaction. *Academy of Management Review*, 8, 501–508.

Tamayo, P., Herrero, A., Martin, J., Navarro, C. & Tranchez, J. (2020). Design of a Chatbot as a distance learning assistant. *Open Praxis*, 12(1), 145–153. https://doi.org/10.5944/openpraxis.12.1.1063.

Winkler, R. & Söllner, M. (2018). Unleashing the potential of Chatbots in education: A state-of-the-art analysis. *Academy of Management Proceedings*, 2018(1), 15903. https://doi.org/10.5465/AMBPP.2018.15903abstract.

World Economic Forum. (n.d.). The fourth industrial revolution, by Klaus Schwab. World Economic Forum. https://www.weforum.org/about/the-fourth-industrial-revolution-by-klaus-schwab.

Zaccone, M. C. & Pedrini, M. (2019). The effects of intrinsic and extrinsic motivation on students' learning effectiveness. Exploring the moderating role of gender. *International Journal of Educational Management,* 33(6), 1381–1394. https://doi.org/10.1108/IJEM-03-2019-0099.

Zawacki-Richter, O., Marín, V., Bond, M. & Gouverneur, F. (2019). Systematic review of research on artificial intelligence applications in higher education — Where are the educators? *International Journal of Educational Technology in Higher Education,* 16(1), 1–27. https://doi.org/10.1186/s41239-019-0171-0.

Chapter 5

Can Gamification Help Undergraduate Students Acquire Leadership and Team-Building Skills? Findings from Two Game Projects

Jayarani Tan, Thomas Menkhoff, Neo Wei Leng,
Bernie Grayson Koh Teck Chye, and Teo Ying Qian Lydia

1. Introduction

With the advent of wireless connectivity and the commoditisation of smart technology, institutions of higher learning need to adapt to the capabilities of the empowered, savvy modern student. In a fight to stay relevant, it is no longer enough for learning activities to be confined within the walls of a classroom. More must be done to reach out and engage students as digital natives in ways that are (more) meaningful to them.

As a remedy to this, blended learning has emerged in recent years as a relevant teaching model (Graham *et al.*, 2013). Commonly understood as a mix of both traditional classroom experiences and online platforms to teach students, blended learning introduces some measure of flexibility to accommodate the unique ways each student understands and assimilates the concepts taught. Delivering blended learning is by no means a 'one-size-fits all' tactic. Educators not only have to master new technology-enabled modes of teaching and learning but they also need to interweave new methodologies and pedagogies within and outside the classroom environment in ways that would best resonate with their students (Moskal *et al.*, 2012).

At the forefront of new teaching and learning technologies used in this context is gamification which refers to the application of game psychology and game mechanics into non-game settings to improve engagement amongst a defined target audience (Hamari *et al.*, 2014). Empirical research on gamification underscores that gamification has had a proven track record in benefitting learning contexts through improved engagement, motivation, and enjoyment. Several studies have shown that games and simulations can foster positive learning outcomes both amongst school going children (Girard *et al.*, 2013) and students of higher learning (Juan *et al.*, 2017; Vlachopoulos & Makri, 2017). It was for this reason that the primary author of this paper chose to adopt gamification during her teaching sections of the Leadership and Team Building (LTB) course, which had been a university core module at Singapore Management University (SMU) since the university was founded in 2000. Until recently, all first-year students, regardless of their specialisation, had to complete this module to meet the requirements for graduation from their chosen field of specialisation (Menkhoff *et al.*, 2018). The LTB course is now an elective embedded into the 'Capabilities' pillar of SMU's new Core Curriculum (the other two pillars are 'Communities' and 'Civilisations') aimed at making students 'competent, adaptable, and responsible decision-makers who can apply themselves to anything, anywhere, for the good of all' (https://corecurriculum.smu.edu.sg/core-curriculum/capabilities/managing).

2. Gamifying Leadership and Team Building in SMU

The overall objective of the LTB module is to build and develop students' competency in leadership and teamwork skills. Students will gain knowledge and skills on leadership and team-building theories, principles, concepts, skills through application, exercises/class activities, self-assessments/instruments, and experiential learning in the form of a group project that entails social innovation.

There several challenges faced in teaching the LTB module. Students do not always see the value or relevance of the module to their chosen field of study and hence are not always enthusiastic about it, sometimes resulting in lower engagement during the term. Another common observation is that students in general are highly concerned about their grades and tend to be driven by grades. Students may also limit themselves to recommended materials for learning with the objective of scoring in their assessments. As leadership and team-building skills bear profound implications in the everyday lives of students, there is a need for greater application of these concepts outside the boundaries of class-based activities.

To address these concerns, experiential group projects were incorporated into the teaching of the LTB module. These projects offer an opportunity for students to work in groups to collaborate with a leader of their choice from a non-profit organisation or a profit-based organisation that upholds community development and/or corporate social responsibility or sustainability initiatives. Students will add value to the needs of the respective beneficiaries that leaders of the organisations will want to impact. Students act as idea champions for social innovation, working in groups, to develop or invent a product or service/programme or put into place a process such as a business plan that will contribute to the organisational leader's goals of impacting community or society at large.

While inclusion of experiential projects was found to be useful, it was noted over the semesters that some students faced challenges in working in teams effectively. The profile of LTB students is diverse in terms of prior knowledge, experience, and even cultural background as there are also several international students. Hence, there is a need to introduce team-building activities which could assist in helping students to familiarise with each other and work more effectively in the group setting. However, the class hours are limiting and team-building activities require time and resources. Thus, to build in opportunities for team building, student engagement, and collaborative and interactive learning, it was decided to harness the potential of technology in the form of a gamified application and use this in addition to the experiential group project and class activities.

The initial outcome of these efforts was the development of a mobile application called 'GameLead,' co-designed with a teaching assistant who eventually went on to set up his own gamification company called Gametize. The project was funded and managed by the university's Centre for Teaching Excellence. Since then, the GameLead application and a follow-up teaching innovation called 'Stranded' have been utilised in the teaching of the LTB module over several semesters, with continual improvement to the application and activities.

Before we share our experiences with these two cases studies and elaborate on how they were designed and how they enhanced teaching and learning, let us take a closer look at the existing topical literature to unearth some of the knowns and unknowns within our topic of choice.

3. Literature Review and Conceptual Framework

Gamification is all about applying game dynamics and frameworks in a non-gaming setting to sustain the interest of the user, such as customers or, as in our case, undergraduate students (Scott & Neustaedter, 2014). The gamification industry first took root in 2010 and has since been on a steady rise. Gamification services are distributed amongst a variety of industry verticals, most notably entertainment, retail, enterprise, and media and publishing (Meloni & Gruener, 2012). It has also been used successfully as a marketing tool by corporations. Gamification has the ability to boost engagement and to improve business operations as of 2011 (Meloni & Gruener, 2012). Corporate powerhouses such as Delta Airlines and IBM2 are but a few of the global organisations that have integrated gamification into employee training for sales and business process management, respectively. Like business-to-employee (B2E) applications, gamification has been found to increase knowledge retention and engagement in popular business-to-consumer (B2C) e-learning platforms such as Khan Academy and Duolingo. Some B2C applications such as Adobe have also jumped on the bandwagon, using gamified elements of progress, achievement, and instantaneous feedback as a fun and easy way of orienting new users to their features.

Over the past few years, gamification has become a game-changer in schools and higher learning institutions (Dichev & Dicheva, 2017; Lee & Hammer, 2011). A gamification platform heightens engagement and enhances motivation while transforming learning into a process of enjoyment and fun as demonstrated by various empirical research studies on this subject matter (Harmari *et al.*, 2014). Student engagement is closely linked to learning effectiveness and outcomes. Research in this domain has established that higher levels of engagement often translate into better learning outcomes, as engaged students are 'good' learners (Handelsman *et al.*, 2005; Gan *et al.*, 2015). The four distinct and reliable dimensions of college student engagement are skills engagement, participation/interaction engagement, emotional engagement, and performance engagement (Handelsman *et al.*, 2005). A learner's level of engagement can vary due to intrinsic factors, or learner variables such as individual motivation, and other extrinsic factors that influence the learning process, such as the involvement and quality of instructors (Jung & Lee, 2018). Ab Rahman *et al.* (2019) examined to what extent gamification enhances the engagement of higher education students by integrating challenges, experience points, levels, badges, and leader boards by comparing data from both gamified and non-gamified courses. Their results suggest that gamification significantly improved students' attention to reference materials, online participation, and proactivity. Gamification also helped learners score better.

In the field of education, gamification can be used to bring about greater commitment, achievement, and involvement in the learning process. Ekici's (2021) systematic review of 22 articles on the use of gamification in flipped learning research suggests that adding game elements into a flipped 'classroom' can lead to higher motivation, participation, and better learning performance. Moodle (a free and open-source learning management system) and Kahoot (a game-based learning platform) turned out to be the most preferred platforms used. Points, badges, and leader boards were the most frequently used game elements. The study by Kasinathan *et al.* (2018) has also revealed positive effects of gamification on students' motivation and engagement compared to traditional teaching modes based on the surveyed case ($n = 24$) of an educational application called Questionify (a 'smart device' developed using C# and Java language) that allows users to collect achievement points as part of their Software Engineering coursework. Respondents felt that gamification can do better in education as compared to the traditional method of teaching students.

Another review of empirical gamification studies regarding higher education and STEM by Ortiz *et al.* (2016) suggests that a combination of game elements (e.g., leader boards, badges, points, and other combinations) has a positive effect on students' performance, attendance, goal orientation, and attitude, e.g., when the subject matter relates to computer science. According to us, there is a general lack of STEM studies with reference to gamification and very little knowledge about which game element is associated with the positive differential impact on student performance, including how mediating/moderating variables such as psychometric measures relate to gamification outcomes. Putz *et al.* (2020) analysed the potential of gamification to foster knowledge retention using action research in the context of a longitudinal study of 617 secondary and tertiary education students conducted over 24 months via various gamified and non-gamified (and constantly refined) workshop designs. Their results suggest that gamification exerts a positive impact on knowledge retention. They also tested the moderating effects of gender and age but found no effect of the former and inconsistent results regarding the latter.

Sanchez *et al.* (2020) extended the theory of gamified learning by focusing on the impact of gamified quizzes on student learning. Their quasi-experimental design study of 473 university students showed that learners who had completed three gamified online quizzes featuring a wager option, a progress bar, and encouraging messages had significantly better scores on the first test compared to students who used traditional quizzes based on a question with four response options. Higher-achieving students derived more benefits from gamification than lower-achieving students. They also argued that there is a 'novelty effect' involved in gamified learning, implying that instructors should not use the same gamification method permanently.

'Traditionalists' have raised noteworthy concerns that the use of game elements such as points, leader boards, and badges may cheapen the motivation to learn amongst students, and that effort will only be applied if the learning activity is deemed 'fun' (Lee & Hammer, 2011). Educational researchers Lee and Hammer (2011) believe that the solution to student disengagement lies with gamification but warn against blind and blanket applications within schools. What seems to be (still) lacking is an understanding of how to design and best apply gamification in formal education vis-à-vis the specific contexts in which it serves to bring the most value to students, instructors, and schools. Similarly, the study conducted by Chan *et al.* (2018) raises a point of caution for educators to be 'very careful' in implementing gamification within the classroom, as there are concerns over how game elements may affect intrinsic motivation, alongside the recommendation for more robust research to analyse the impact of gamification in education.

A useful theoretical model to design 'good' gamified teaching and learning strategies for formal education is 'The Framework for the Rational Analysis of Mobile Education' (FRAME) developed by Koole (2009). The term 'mobile' implies the use of mobile devices such as personal electronic devices (tablets, laptops, and smart phones). FRAME comprises the device, the learner, and social aspects that influence and co-produce learning. In the FRAME Venn diagram, each circle is of the same size symbolising that they are equally important.

Another 'practical' gamification framework is 'The Octalysis Framework,' a human-centric gamification design framework developed by Chou (2015) that features eight 'core drives' for human motivation: Epic Meaning and Calling, Development and

Accomplishment, Empowerment of Creativity and Feedback, Ownership and Possession, Social Influence and Relatedness, Scarcity and Impatience, Unpredictability and Curiosity, and Loss and Avoidance.

Garcia-Penalvo's (2021) TPACK model draws attention to the importance of technological knowledge (TK), pedagogical knowledge (PK), and content knowledge (CK) as constituting components of a robust e-learning approach. It provides useful checks so that the chosen technology, pedagogical (student-centred) approach, and content knowledge (e.g., effective teamwork) are effectively linked to the competencies (learning outcomes) students are expected to acquire by the end of a particular class/course or programme so that they appreciate why those competencies are useful to them. Student-centred teaching and learning implies that instructor and students co-construct knowledge, e.g., by making them work in teams where they can put their strengths to use by relying on each other to leverage each other's knowledge to achieve something greater than what everyone would have achieved alone. In such a context, the instructor is more like a facilitator who must select the 'right' approach to engage students such as case-based learning, problem-based learning, project-based learning, or as in our case a hybrid (mixed) approach of problem-based, gamified learning centred on videos and in-class debriefings. All these approaches are structured differently and require different instructional designs.

The three theoretical frameworks described earlier point to important variables that influence the effectiveness of gamification as an engagement tool in non-gaming contexts such as education. They provide guidance for the analysis of our two case studies (Eisenhardt & Graebner, 2007), 'GameLead' and 'Stranded,' which will be introduced in the following to provide the reader with a better understanding of successful gaming mechanics in educational courses.

4. Case Study Method

To contribute to the still nascent body of knowledge pertaining to effective gaming mechanics in educational contexts, this paper presents two case studies of innovative gamification applications designed by the first author: 'Gamelead' and 'Stranded.' An explanatory, descriptive, and exploratory case study approach (Gerring, 2007) helps to analyse the specific issue we are interested in, namely, assessing the potential of gamification to engage learners so that they achieve the desired learning outcomes, within the boundaries of a specific course, i.e., in our case Leadership and Team Building (LTB). The descriptive–analytical comments as well as the lessons learned are based on several forms of evaluative data collected through post-course feedback surveys, in-class observations, and informal after-action reviews conducted by us. Due to the lack of sufficient quantitative data, we are unable to provide real evidence about the imputed effect of the two gamified learning contexts on students' engagement. Rather, the emphasis is on sharing why and how we developed and deployed these gamified applications within the specific context of the Leadership and Team Building course aimed at imparting relevant competencies into students such as team leadership and providing generic recommendations for other instructors interested in implementing gamification solutions in their classes. Just as leaders in organisations blend traditional and new skills to steer their organisations into the future (Kane *et al.*, 2019), educators need to blend traditional and new teaching skills to prepare their students for the digital world of work. The case study approach is useful to revisit and further enhance relevant theoretical concepts as we will demonstrate in the following.

5. Results: Case Studies and Assessment

5.1. *Case Study 1: GameLead Application*

'GameLead' is a mobile and desktop application that is compatible with android and apple devices. It was introduced as part of the LTB learning process in the classroom to enhance achievement beyond textbook knowledge and classroom learning. The application presents students with weekly quests of 8–9 challenges on content related to leadership and team building over a period

Figure 1.   GameLead Introductory Orientation Screens

of 10 weeks. The content posted comprises quizzes, reflective questions, and discussion of article and/or videos related to the LTB subject matter. Students choose and swipe to start and finish with all the quests posted and record their inputs in the application. To help the students in using the application, they were given a set of orientation screens (see Figure 1) at the start of the first lesson.

Subsequently, weekly quests were made available at the beginning of the week for students to participate in, with preceding quests being left unlocked for them to review. Each week's quest contained a series of challenges that prompted students to reflect, apply, and act upon ideas that were brought up during the class for that week. Examples of challenges included watching videos on exemplary leaders, snapping photos of team activities, answering simple quiz questions, and reflecting on and sharing of thoughts and insights. The challenges were designed with the objective of piquing the interest of the students, making them relate the theoretical content of the class to real-life contexts, helping them think critically, get to know their team members, and work collaboratively with their team, and encouraging them to learn on their own and from their peers (see Figure 2).

Figure 2.   Examples of GameLead Challenges

Each week's 'GameLead' inputs would be discussed by students the following week during class with a debrief summarising the lessons learnt. The TA is assigned with the task of facilitating the discussion session of 'GameLead' activities, which is a non-graded component. Students compete at a group level, and the winning group at the end of the seven weeks of GameLead would get to choose the order of their final oral presentation for their group project.

Upon completing a challenge, students would receive points within 'GameLead.' With these earned points, students could choose their respective group presentation slots as a sort of incentive to gain from convenience and flexibility of choice. The application interface also included social features such as a newsfeed section, where students could view, vote, and comment on the past submissions of their classmates. These social features do promote in-app interaction and drive student engagement through collaborative and interactive learning from their peers. To introduce a degree of competition within the app, students were ranked accordingly on a live leader board based on the number of points they had received by completing the given challenges.

The application was used by the students outside of class, and their online reflections and discussions (using the application) were integrated into the in-class sessions through weekly discussions. It should be noted that the use of 'GameLead' as part of the course is voluntary and not graded. Yet, all students participated actively.

As 'GameLead' increased student participation and engagement immensely compared to traditional methods of teaching without technology (Tan & Sockalingam, 2015), SMU's Centre for Teaching Excellence (CTE) decided to make this app available to all instructors who might be interested in using it for their own teaching purposes. Instructors could change the contents by replacing the original with theirs to suit the subject matter and the learning outcomes as deemed fit for their own courses.

5.2. *Evaluation of GameLead Application*

To understand the impact of the GameLead application on student engagement, an online survey was conducted at the end of the course in the second semester of academic year 2020/2021. The quantitative questions queried on students' perceptions of the learning as well as cognitive and affective engagement. The qualitative questions explored support needed from the instructor, design of activities, positive experiences, and suggestions for improving the application.

A total of 119 students from two courses responded on the survey. Using a 7-point Likert scale, the results show a statistically significant increase in perceived knowledge after using 'GameLead' ('Somewhat High,' $M = 5.18$, SD = 1.06) compared with before using it ('Somewhat Low,' $M = 3.35$, SD = 1.37).

In terms of learning, 88% of respondents found that the application had delivered the contents in a clear manner, and this was attributed to the clear instructions given. About 85% of respondents indicated that the game experience enhanced their ability to make connections to real-life issues. Most (87%) respondents agreed that 'GameLead' allowed them to learn at their own pace.

About two-third (73%) of the respondents enjoyed the challenges, and 72% reported that they completed the tasks even without rewards (even though 52% found the tasks challenging).

Taken together, these results are suggestive that students are intrinsically motivated to use this application for their learning despite the challenges of the activities and learning associated with it.

To understand more, qualitative comments from the survey were analysed for commonly occurring themes. Students cited the use of interesting, relevant videos and the questions following the videos to be engaging. They found activities requiring reading of others' posts and voting or commenting to be useful in learning from their peers and in guiding them to consider alternative perspectives. Besides the interactive learning experience, the competition helped motivate the teams to generate and showcase more ideas during the 'GameLead' activities. One student commented, "Gamification helps to recap learning" (S9) and another said, "It allows me to review the content at my own free time. This flexibility makes learning more enjoyable, hence I actually studied more which help me understand the concepts" (S18).

On the other hand, students felt that more could be done to integrate the online discussions and activities on the application with the face-to-face classroom learning for that week. They also suggested more challenging and varied types of questions to make the learning experience even more engaging. To enhance the usability features of the application, they requested the ability to retain the previous weeks' answers for revision. Student feedback also suggested that the team needs to rethink and develop a better reward scheme.

Overall, the results indicated that the objectives of the 'GameLead' application to create opportunities for team building, student engagement, and collaborative and interactive learning had been met beyond expectation. Even though students did not get any additional grades for using the application, and the rewards were only appealing to half of the respondents, most students found the application to be useful for the intended purposes.

5.3. *Case Study 2: Stranded Application*

The digital game 'Stranded: A Spark of Hope' ('Stranded' in short) was developed by SMU's Centre for Teaching Excellence (CTE) and the Singapore University of Technology (SUTD) Game Lab for classroom use. This was done not only to engage students but also to see how well students can apply their concepts and theories to engage in critical problem solving in a simulated situation by way of a virtual game. It took about a year to develop the gaming platform, which was subjected to several rounds of testing with students to overcome and address any technical glitches and fine-tune other usability functions of the digital game.

The game which has been in use since 2016 requires about 40–60 minutes of play time during classroom lesson time. The storyline of the game revolves around a van that breaks down with four people travelling together in it who appear to be stranded in enemy territory (see Figure 3). The delivery mode of the game is digital and is intended to be played on a computer or a mobile device. The strategic intent takes the form of an immersive approach, where the player takes on the role of a leader.

The player must garrison his/her team members, comprising four followers with different traits and proficiencies, namely, Jessa (the only female; her persona characterises her as aloof and intolerant), Jiro (assertive, opinionated, and stoic), Bastian (amenable and protective), and Daya (supportive and zealous), to help fix the van and survive with food and shelter till they leave the territory as soon as possible. The task of getting the van working again is the only rescue plan or else they remain stranded in an unfriendly, foreign territory and are left to face enemy attack for trespassing.

In the absence of animation, owing to budget constraints, the flow of the game relies on conversations and communication that prompt (scored) action via task assignments by the leader such as scavenging (which jeopardises the ability to defend the camp) or to repairing the vehicle with the scavenged materials. The depictions in Figure 4 illustrate an example of the personas engaged in conversations.

Figure 3. Broken Down Van That Needs Fixing

At the end of each task attempt, scores are provided as to how the student fared as a leader, indicating his/her leadership capabilities and shortcomings obtained from the feedback analytics as shown in Figure 5. A successful outcome would depend on the knowledge, skill, and ability of the leader to lead the team out of the crisis. The game ends by showing participants an analytics page, depicting the player's leadership aptitudes as illustrated in Figure 5.

The final score depicting an 'S' indicates an outstanding or excellent attempt made on the part of the leader while 'A' which denotes a very good attempt. 'B' denotes good attempt and 'C' shows a 'satisfactory' attempt. A successful attempt would mean that the leader has demonstrated his/her ability to lead the team out of danger as shown in Figure 6.

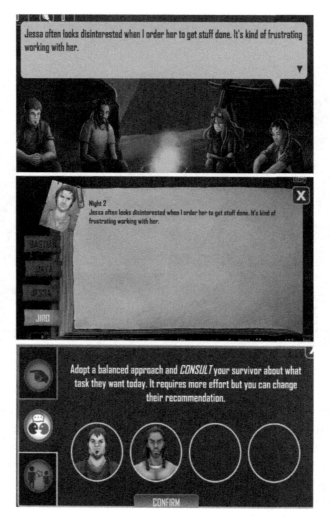

Figure 4. Examples of Communication Taking Place in the Team

Despite the fact that no incentives were provided, the game proved to be engaging as students were absorbed in wanting to be able to succeed.

The game development process involved continuous meetings with the instructor and discussions with the SUTD Game Lab game developers and engineering architects, who were responsible for the

Figure 5. Leadership Analytics Shown at the End of the Game

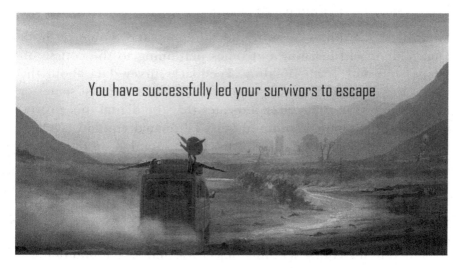

Figure 6. Successful Outcome of Leadership and Teamwork

mechanics of game creation. This allowed the instructor to explain what the learning aspects of Fiedler's contingency theory of leadership, Adair's action-centred leadership, Lewin's group dynamics concept, and Thomas/Kilmann's conflict management approach needed to be incorporated into the game so that it would be an effective teaching tool (Lameras *et al.*, 2017).

Prior to introducing the game at the classroom level, several tests were conducted with the instructor's teaching assistants, who were senior students. The initial feedback from the teaching assistants showed that while the level of usability of the various functions of the game was favourable, the level of playability proved to be too much of a challenge. As such, adjustments were made progressively at the end of each of the three rounds of such testing.

Testing was also done at the classroom levels across the four sections of undergraduate students with each class comprising about 45 undergraduate students. The game developers were present in class to solicit feedback from the students to see if any further refinement of the game was needed to improve the playability level. Finally, the game was launched in 2016 and it has been in use since then.

5.4. *Evaluation of Stranded Application*

'Stranded' continues to intrigue students and enhance both engagement and learning outcomes pertaining to the course on Leadership and Team Building. Feedback survey data about the use of the 'Stranded' game collected by CTE in the first semester of academic year 2020/2021 indicates that the digital tool proved to enhance student learning about LTB-related content, such as team leadership. A total of 143 students responded on the survey. Using a 7-point Likert scale, the results show a statistically significant increase in perceived knowledge after using 'Stranded' ('Somewhat High,' $M = 5.51$, SD $= 0.87$) versus before using it ('Average,' $M = 4.38$, SD $= 1.21$).

In terms of learning, 96% of respondents agreed that 'Stranded' is aligned to the course learning objectives, and 92% found that the application had delivered the contents in a clear manner. About 94% of respondents indicated that the game experience enhanced their ability to make connections to real-life issues. A majority (91%) of the respondents agreed that 'Stranded' allowed them to learn at their own pace and 88% felt that it provided them with timely feedback for their learning. As for engagement, 90% of respondents enjoyed their learning and 91% agreed that 'Stranded' had stimulated their curiosity about the topic of leadership and team building.

The data about the use of the 'Stranded' game collected by CTE indicate that the game proved to enhance student learning about LTB-related content. As a suggestion for improvement, learners expressed the desire for more instructions on how to play the game successfully. Students also wished for more variety in terms of the tasks, people's attributes, concepts, and examples.

The content of the game, 'Stranded: A Spark of Hope,' and how it is to be employed by any instructor has been made available since 2019 through the Case Centre at SMU as well as Harvard Business Publishing.

Students appraised the game favourably in terms of learning outcomes. Table 1 specifies the learning outcomes of the 'GameLead' and 'Stranded' games.

Table 1. Bloom's Learning Objectives and Outcomes of 'GameLead' and 'Stranded' Games

Bloom's Learning Objectives	'GameLead' Learning Outcomes	'Stranded' Learning Outcomes
Create	Through the application and the weekly reflective challenges, students learn to think creatively which helped them create a better understanding of their own LTB skills and those of their team members.	Through the application and the various scored task attempts to get to safety, students learn the importance of adapting one's leadership style to changing situations and people contexts. Feedback analytics create deeper insights into one's own leadership capabilities, including those related to development needs and the way forward.

(Continued)

Table 1. (*Continued*)

Bloom's Learning Objectives	'GameLead' Learning Outcomes	'Stranded' Learning Outcomes
Evaluate	The 'GameLead' platform helps students practice their LTB knowledge acquired from the class and readings by appreciating LTB concepts and honing their collaborative problem-solving skills.	The 'Stranded' application helps students practice their LTB knowledge acquired from the class and readings by appreciating leadership theories and honing their interpersonal task achievement skills, related to communication and conflict management, and other people skills.
Analyse	Through several reflective and team-building activities, students learn to break down information into component parts, connect the dots, and solve challenges in a virtual community space.	Students tackle several LTB challenges related to a crisis situation such as being stranded in a desert. They learn to deal with the situational complexities and the different follower personalities (including their unique behavioural traits) while working as a team to achieve a common goal.
Apply	Students apply relevant LTB concepts, accurately assess their own strengths and weaknesses as leaders, and develop a personal plan for leadership development.	Students employ an appropriate leadership style and decision-making approach for each game day given the situational challenge and people context to accomplish the end goal of the simulation exercise.
Understand	Students understand how concepts and theories introduced in the classroom in a fun manner relate to them as leader and team player.	Students understand what it means to be an effective team leader managing a crisis and working towards a viable solution with the involvement of team members.

Table 1. (Continued)

Bloom's Learning Objectives	'GameLead' Learning Outcomes	'Stranded' Learning Outcomes
Remember	Through the weekly attempts on the app, students deepen their knowledge and understanding of both leadership and teamwork concepts and are able to use them beyond the confines of classroom learning.	Through the 'Stranded' game, students reinforce their understanding of LTB concepts related to situational leadership styles, decision-making, communication, and conflict management.

6. Discussion

Considering how both applications were used in the LTB module and students' positive feedback, we postulate that one possible reason for successful deployment in the LTB module is the structured use of the application. For instance, the 'GameLead' application involved weekly quests of 8–9 challenges on a regular basis. Such regularity in use sets clear expectations and provides continuity for learners. Also, the instructor and teaching assistants made an effort to properly introduce the application and regularly brief students about the weekly participation. This helped blend the online and face-to-face learning seamlessly. Contrastingly, if students were left to their own means to use the application without integrating it into classroom teaching, it is likely that the application may not have been that useful. Thus, educators planning to use gamified applications in teaching and learning are well advised to effectively design them as part of an engaging, blended learning process.

Students' post-course evaluation responses indicate that the success of the 'GameLead' and 'Stranded' applications in terms of acquiring LTB-related content can be attributed to three critical engagement factors as outlined in Koole's (2009) FRAME model: (1) technological factors, (2) pedagogical factors, and (3) sociological factors. Technological factors in our case include the ease of using the tool and the user interface of the application, while pedagogical factors refer to the instructional value of the application such as relevance, level of challenge, and extent of authenticity. Sociological factors include opportunities created for the students to bond with their team, to interact with others, and to participate in collaborative learning activities through the use of the tool. Students who used the 'GameLead' and 'Stranded' applications gave high ratings for all three factors, with some suggestions for continual improvements as shared in the following.

Comments by learners about potential enhancements of the applications underscore the relevancy of the Octalysis Gamification Framework and its eight categories. Students' feedback suggests that 'rewards' represent an important area for improvement as only 50% were enticed by the rewards used in the application. While student engagement was high despite this, it would be essential to address the other elements of gamification to improve student engagement.

Overall, the 'GameLead' application was rated to be high on two factors of the framework: (1) epic meaning (novelty and impactful) and calling, and (2) social influence and relatedness (connecting mentally and emotionally with the game situation). This indicates that in future iterations, it would be important to consider integrating more gaming elements into 'GameLead' which would meet other component requirements of the Octalysis framework such as empowerment, scarcity, avoidance, accomplishment, ownership, and unpredictability to make the user experience (even) more engaging and meaningful. Several of these factors (e.g., scarcity and accomplishment) are key elements of the 'Stranded' application which may explain its success.

If we relate the above back to the theory and practice of student engagement in the gamified LTB classroom, we can conclude that a learner's level of engagement can vary due to intrinsic factors such as one's individual (intrinsic) motivation and extrinsic influence factors such as the involvement and quality of instructors, mentors, and feedback opportunities as well as the overall quality of the game (Monteiro *et al.*, 2015; Leng *et al.*, 2010). Student engagement is closely linked to learning effectiveness and outcomes. Research in this domain has established that higher levels of engagement often

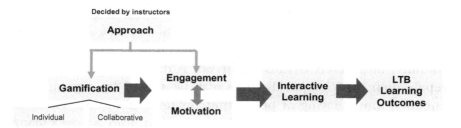

Figure 7.   Engaging Undergraduate Students through Gamification: A Model

translate into better learning outcomes (Handelsman *et al.*, 2005). Figure 7 depicts the logic and the way in which the two case studies work according to our analysis to drive students' engagement and motivation through gamification aimed at achieving desired LTB learning outcomes. This model could serve as conceptual basis for a future representative study.

7. Conclusion and Lessons Learned

In this chapter, we introduced and contrasted two gamified teaching and learning applications that we are using in a Leadership and Team Building core course at an Asian university: (i) 'GameLead,' which features collaborative learning mechanics, and (ii) 'Stranded,' a student-centred, self-directed learning game where each individual takes on the role of a team leader.

The 'GameLead' application, accessible on mobile and desktop platforms, presents students with various weekly challenges on topics relevant to leadership and team building. By participating in these weekly challenges at their own convenience, students earn points, collaborate with peers, and are rewarded with special privileges. They also compete against each other via a leader board. Student participation and the earned points from the 'GameLead' application are integrated into the face-to-face lessons.

'Stranded' is designed as a simulation game that provides an experiential platform to acquire situational team leadership competencies and interpersonal skills based on sudden changes of events.

The setting is a future dystopian world where resources are scarce and different cities are at war with each other as they compete for resources. Players take on the role of a leader whose team has been stranded in a desert because their van is damaged. If they fail to repair their vehicle quickly, they may be raided by surrounding enemies, thus delaying or preventing any hope of escape. The aim of the game is to enable each individual player, who is now a team leader, to grapple with the complexities of the given situation, including the team members' varied personalities, so that the team gets out of the present predicament.

From the perspective of Bloom's hierarchy (Bloom, 1956), the 'Stranded' game shows a higher-order complexity because it is more challenging as students are forced to try out different things to achieve their goals. Table 2 compares the 'Gamelead' and 'Stranded' games.

Overall, our research results suggests that the successful design and deployment of gamified applications in education can be attributed to well-planned, pedagogy-informed constructivist

Table 2. Comparison of 'GameLead' and 'Stranded' Games

Comparison Components	'GameLead'	'Stranded'
Goals	Assignment-based learning with the aim of deepening knowledge beyond classroom learning. Groups in class compete.	Action driven and experiential decision-making by immersion in game situation at individual level.
User Engagement	Group collaboration with fun elements.	Individuals are thrilled to be able to tackle challenges during a crisis.
Challenge	Pressure to not let the group down in the competition.	Prove one's own leadership capability by applying respective knowledge, skills, and abilities.
Achievement	Teamwork collaboration.	Resilience and maturity to make independent decisions.
Motivation	Winning at class level.	Learning more about oneself.

Table 2. (*Continued*)

Comparison Components	'GameLead'	'Stranded'
Learning	Collaborative learning.	Student centred and self-directed.
Competition	Striving to win at class level by establishing superiority over classmates.	Striving to overcome personal challenges.
Leader boards showing the names and scores of others (e.g., competitors)	Class level.	Personal level only.

teaching and learning approaches that harness the potential of technology such as gamified mobile platforms that motivate and engage learners. As shown in this study, gamification can be successfully coupled with blended and mobile learning in institutions of higher education to engage learners. Koole's Framework for the Rational Analysis of Mobile Education, the Octalysis Gamification Framework, Garcia-Penalvo's (2021) TPACK model, and the theory of student engagement can help instructors keen to implement gamification effectively design such novel applications as well as decipher possible ways to further improve them. Our experiences demystify the notion that gamification will only work for younger children by adding some case-based materials to the growing body of gamification knowledge pertaining to the smart (smarter) use of game mechanics in higher education (and possibly adult learning).

As Sanchez *et al.* (2020) have stressed, there is often a 'novelty effect' at work when it comes to successfully rolling out gamified teaching and learning approaches. They also warn that instructors should not use the same gamification method permanently. While we also noticed the novelty effect in our classes, we would argue that the underpinning pedagogy is indeed very critical to engage learners over time. It is essential that the challenges, activities, and rewards are carefully planned so that engagement does not merely stem from 'fun' or 'socialising' activities. The effective design of gamified applications requires a good understanding of what sort of activities drive team-building and collaborative exercises towards the desired learning outcomes. This study helps to corroborate our conceptual ideas and gamification model (see Figure 7), which can serve as a conceptual basis for a future representative, empirical study on the topic.

One limitation of our study is that the application was only used in one particular course with thematic focus on leadership and team building. There is a need to test it further in other modules and collaborative learning contexts. Another limitation is that our study involved only students' perceptions of their learning. In future studies, we plan to collate additional indirect measures such as students' responses/comments on peer's posts to study interactive learning patterns and to correlate the use of the application with learning outcomes (for instance, by analysing the quality of work submitted by students through 'GameLead'). We also plan to study the impact of different types of game elements such as plot, tension, conflict, resolution, and strategy on student engagement and learning.

Availability of Data and Materials

The datasets generated and/or analysed during the current study are not publicly available due to limitations set by the Institutional Review Board (IRB) at Singapore Management University but are available from the corresponding author on reasonable request.

Funding

Both applications 'GameLead' and 'Stranded' were funded by the Technology Enhanced Learning Fund (TEL Fund) borne by the Centre of Teaching Excellence at Singapore Management University.

References

Ab Rahman, R., Ahmad, S. & Hashim, U. R. (2019). A study on gamification for higher education students' engagement towards education 4.0. In V. Piuri, Balas, V. E. Borah, S. & S. S. Syed Ahmad (Eds.), *Intelligent and Interactive Computing* (pp. 491–502). New York: Springer.

Bloom, B. S. (1956). *Taxonomy of Educational Objectives: The Classification of Educational Goals.* New York: David McKay.

Chan, E., Nah, F. F.-H., Liu, Q. & Lu, Z. (2018). Effect of gamification on intrinsic motivation. In *International Conference on HCI in Business, Government, and Organizations. HCIBGO 2018: HCI in Business, Government, and Organizations*, pp. 445–454. https://doi.org/10.1007/978-3-319-91716-0_35.

Chou, Y.-K. (2015). *Actionable Gamification: Beyond Points, Badges and Leaderboards.* Singapore: Createspace Independent Publishing Platform.

Dichev, C. & Dicheva, D. (2017). What is known, what is believed and what remains: A critical review. *International Journal of Educational Technology in Higher Education*, 14(9). https://doi.org/10.1186/s41239-017-0042-5.

Eisenhardt, K. M. & Graebner, M.E. (2007). Theory building from cases: Opportunities and challenges. *Academy of Management Journal*, 50(1), 25–32. https://doi.org/10.5465/AMJ.2007.

Ekici, M. (2021). A systematic review of the use of gamification in flipped learning. *Education and Information Technologies*, 26, 3327–3346. 10.1007/s10639-020-10394-y.

García-Peñalvo, F. J. (2021). Avoiding the dark side of digital transformation in teaching. An institutional reference framework for eLearning in higher education. *Sustainability*, 13(4), 2023. https://doi.org/10.3390/su13042023.

Gan, B., Menkhoff, T. & Smith, R. (2015). Enhancing students' learning process through interactive digital media: New opportunities for collaborative learning. *Computers in Human Behavior*, 51, 652–663.

Gerring, J. (2007). *Case Study Research: Principles and Practices.* Cambridge: Cambridge University Press.

Girard, C., Ecalle, J. & Magnan, A. (2013). Serious games as new educational tools: How effective are they? A meta-analysis of recent studies. *Journal of Computer Assisted Learning*, 29(3), 207–219.

Graham, C. R., Woodfield, W. & Harrison, Buckley, J. (2013). A framework for institutional adoption and implementation of blended learning in higher education. *The Internet and Higher Education*, 18, 4–14. doi:10.1016/j.iheduc.2012.09.003. ISSN 1096-7516.

Hamari, J., Kolvisto, J. & Sarsa, H. (2014). Does gamification work? — A literature review of empirical studies on gamification. In *Proceedings of the 47th Hawaii International Conference on Systems Sciences*, Hawaii, USA, January 6–9.

Handelsman, M. M., Brigss, W. L., Sullivan, N. & Towler, A. (2005). A measure of college student course engagement. *The Journal of Educational Research*, 98(3), 184–191.

Juan, A. A., Loch, B., Daradoumis, T. & Ventura, S. (2017). Games and simulation in higher education. *International Journal of Educational Technology in Higher Education*, 14(7), 1–3.

Kane, G., Philps, A. N., Copulsky, J. & Andrus, G. (2019). How digital leadership is(n't) different. *MIT Sloan Management Review*, 60(3). REPRINT # 60309.

Kasinathan, V., Mustapha, A., Fauzi, R. & Rani, M. F. C. A. (2018). Questionify: gamification in education. *International Journal of Integrated Engineering*, 10(6). https://publisher.uthm.edu.my/ojs/index.php/ijie/article/view/2781.

Koole, M. L. (2009). A model for framing mobile learning. In M. Ally (ed.), *Mobile Learning: Transforming the Delivery of Education and Training*. Athabasca, AB: Athabasca University Press.

Lameras, S. A., Arnab, I. D., Stewart, S., Clarke, S. & Petridis, P. (2017). Essential features of serious games design in higher education: Linking learning attributes to game mechanics. *British Educational Research Association*, 48(4), 972–994.

Lee, J. & Hammer, J. (2011). Gamification in education: What, how and why bother? *Academic Exchange Quarterly*, 15(2), 1–5. https://www.researchgate.net/publication/258697764_Gamification_in_Education_What_How_Why_Bother

Mak, Heong Weng (2013). *The Gamification of College Lectures*. University of Michigan. http://tinyurl.com/brwxvax.

Meloni, W. & Gruener, W. (2012). Consumer and enterprise market trends, M2 research. https://www.docsity.com/pt/gamification-in-2012/4799762/.

Menkhoff, T., Tan, R., Kan, S. N. & Tai K. (2018). Integrating social innovation into a leadership and team building course. In T. Menkhoff, K. S. Ning, H.-D. Evers & C. Y. Wah (Eds.), *Living in Smart Cities: Innovation and Sustainability* (pp. 303–325). Singapore: World Scientific Publishing.

Monteiro, V., Mata, L. & Peixoto, F. (2015). Intrinsic motivation inventory: Psychometric properties in the context of first language and mathematics learning. *Psicologia, Reflexão e Crítica*, 28(3), 434–443. https://doi.org/10.1590/1678-7153.201528302.

Moskal, P., Dziuban, C. & Hartman, J. (2012). Blended learning: A dangerous idea? *Internet and Higher Education*, 18, 15–23. doi:10.1016/ j.iheduc.2012.12.001.

Ortiz Rojas, M. E., Chiluiza, K. & Valcke, M. (2016). Gamification in higher education and STEM: A systematic review of literature. In *8th International Conference on Education and New Learning Technologies (EDULEARN)* (pp. 6548–6558). IATED Int Assoc Technology Education & Development. DOI: 10.21125/edulearn.2016.0422.

Putz, L. M., Hofbauer, F. & Treiblmaier, H. (2020). Can gamification help to improve education? Findings from a longitudinal study. *Computers in Human Behavior*, 110, 106392. https://doi.org/10.1016/ j.chb.2020.106392.

Sanchez, Langer, M. & Kaur, R. (2020). Gamification in the classroom: Examining the impact of gamified quizzes on student learning. *Computers and Education*, 144, 103666–. https://doi.org/10.1016/ j.compedu.2019.103666.

Schaaf, R. & Quinn, J. (n.d.). 12 examples of gamification in the classroom. https://www.teachthought.com/the-future-of-learning/12-examples-of-gamification-in-the-classroom/.

Scott, A. & Neustaedter, C. (2014). Analysis of gamification in education. School of Interactive Arts and Technology, Simon Fraser University. http://clab.iat.sfu.ca/pubs/Stott-Gamification.pdf.

Selawsky, J. (2019). Gamification in education: A new type of interactive learning. https://elearningindustry.com/gamification-in-education-new-type-interactive-learning.

Tan, R. & Sockalingam, N. (2015). Gamification to engage students in higher education. Working Paper, Research Collection Lee Kong Chian School of Business, pp. 1–14.

Vlachopoulos, D. & Makri, A. (2017). The effect of games and simulations on higher education: A systematic literature review. *International Journal of Educational Technology in Higher Education*, 14(22). https:// doi.org/10.1186/s41239-017-0062-1.

https://doi.org/10.1142/9789811293108_0006

Chapter 6

Sensitising Students in Higher Education to Appreciate the Importance of Digital Sustainability for Smart Cities

Thomas Menkhoff and Kan Siew Ning

1. Background and Course Context

The 21st century has been described as the urban century in view of rapid urbanisation rates that have led to the growth of megacities with populations of over ten million people. Many of these large cities are located in the developing world, often with an inverse relationship between economic growth and urbanisation despite the potential of cities as engines of growth and development. All big cities such as New York, London, Tokyo, Paris, Shanghai, Hong Kong, Singapore, New Delhi, and Jakarta, must cope with major urban challenges in areas such as climate change, high population density, energy, poverty, health, access to basic utilities, pollution (air and ground), housing, and transport. (Kumninos, 2009; Paskaleva, 2009; Menkhoff *et al.*, 2018; Foo & Pan, 2018; Wang & Kiat, 2018; Miller, 2018; Phang, 2018)

How do we create sustainable cities which are spacious, green, connected, fair, and safe? How can the concept of urban sustainability be deployed to engender 'good' cities which are inclusive, accessible, and effectively governed? How can the economic potential and performance of 'global cities' be enhanced to create local economic development in support of the 2030 Agenda on Sustainable Development (especially SDG 11 — making cities inclusive, safe, resilient, and sustainable)? How can urban planning, the appropriate design of the physical environment, and novel technical 'smart city' solutions (e.g., electronic data collection sensors which supply information that can be used to better manage urban flows and allow for real-time responses) be utilised to tackle urban challenges? What does it take to create real urban communities where people can 'live, play, and work' well? Is a 'digital city' automatically a 'smart' city? Is a smart city also sustainable? What does 'sustainable' really mean?

The course 'SMT112 Sustainable Digital City' aims to provide answers to these questions with special emphasis on the managerial aspects of urban sustainability and related 'smart' city applications (Sassen, 2005; Williams, 2009; Keivani, 2009; The Straits Times, 2018; Kong, 2018). Starting from the stakeholder requirements of people, citizens, city managers, and planners of sustainable and innovative cities, the course is aimed at enabling students to explain what a sustainable city is and how to bring about sustainable urban development in a 'smart,' digital city context.

With the help of case studies and resource persons such as industry leaders, planners and innovative city designers, technology experts, and business development experts from local and international organisations, students are familiarised with the opportunities and challenges of creating sustainable digital cities vis-à-vis the smart city thrusts of the Singapore Government and related businesses. Local site visits of 'smart' urban sites in Singapore complement the learning experience.

1.1. *Learning Objectives*

The overall objective of this module is to equip students with core knowledge of what it takes to plan, design, build, and sustain digital, 'smart' cities that are innovative, inclusive, safe, resilient, and sustainable.

By the end of this course, it is expected that students appreciate the following 4 areas:

Urban Challenges
- Explain why climate change, high population density, energy, poverty, health, access to basic utilities, pollution (air and ground), housing, transport, etc., represent challenges for cities in the North and the South.

Concept of Urban Sustainability and Taxonomy of Sustainable ('Smart') Cities
- Explain how city leaders and planners can deploy the concept of urban sustainability to create sustainable cities which are spacious, green, connected, fair, safe, inclusive, accessible, and effectively governed.
- Appreciate the need to enhance the economic potential and performance of 'global cities' to create local economic development in support of the 2030 Agenda on Sustainable Development.
- Describe the core characteristics of a 'Smart City' and how smart city components can add value in terms of innovativeness, livability, and sustainability.

Design and Functionality of 'Smart' Cities
- Explain how urban planning, the appropriate design of the physical environment, and novel technical 'smart city' solutions such as electronic data collection sensors or smart grids can be utilised to tackle urban challenges and to make a city 'smart(er).'
- Appreciate the challenges in creating robust (e.g., inclusive) communities where people can 'live, play, and work' well.
- Explain what it takes to ensure that a 'digital city' is automatically a 'smart' and 'sustainable' city.

Commercialisation Aspects of the Sustainable Digital City Concept
- Appreciate the challenges in successfully commercialising digital ('smart') city concepts and applications amidst increasing competition in this field.

- Know some of the key players in the local and international context that participate in this service sector and establish network contacts.

1.2. *Teaching and Learning Contents*

The course 'SMT112 Sustainable Digital City' is one of several courses of the Smart-City Management & Technology Programme offered as one of 4 tracks within the university's Information Systems major. This focus area is part of the Bachelor of Science (Information Systems) curriculum of the SMU School of Computing and Information Systems. 'SMT112 Sustainable Digital City' comprises of ten different topics ranging from Singapore's early sustainability challenges to smart building technology R&D. During the first session, students are introduced to the evolution of the sustainability movement at international levels, followed by a deep dive into the Singapore story and the efforts of Singapore's founding fathers to develop Singapore (from a Third World country to First World country) with a view towards appreciating the city's physical transformation under the visionary leadership of former PM Lee Kuan Yew. Special emphasis is placed on the concept of 'urban sustainability,' why it matters, and associated challenges.

To appreciate the concept of 'sustainability,' students visit PUB's 'Sustainable Singapore Gallery' (https://www.pub.gov.sg/marinabarrage/ssg) which showcases both the issues pertaining to sustainability as well as the policy measures adopted by the Singapore Government to manage it such as mitigating the effects of climate change or managing e-waste. The visit helps students understand the importance of technology in dealing with urban challenges such as water scarcity, global warming, pollution, or waste management. A major takeaway from this first site visit is the insight into how fragile Singapore is, e.g., in terms of water, climate, waste, and energy, and that there are indeed several digital solutions that can help tackle urban problems ranging from flood control to smart grids.

The third session sheds light on the role of (digitally enabled) urban planning in the context of 'good' urban governance and urban sustainability. If feasible, a visit to URA's new digital planning lab complements the session to introduce students to URA's new digital tools "that bring together all available data on a single platform for advanced analysis and visualisation. The ability to overlay, merge and model different layers of data makes it possible to quickly and easily uncover new relationships and patterns. These can help to reduce uncertainty and 'what if' elements of the urban planning equation" (https://www.ura.gov.sg/Corporate/Resources/Ideas-and-Trends/Harnessing-the-Power-of-Digital-Technologies). An example of such as tool is the Geospatial Information System (GIS)-Enabled Mapping Modelling and Analysis (GEMMA) decision support tool, which enables planners to conduct integrated planning analytics and simulation. The tool was developed in collaboration with the Institute for Infocomm Research (I2R) and other planning agencies.

In the fourth week, students are introduced to the importance of having a stable infrastructure as the basic building block of a sustainable city. Starting from how Singapore built labour-intensive industries in the 1970s and later progressed to stock exchanges and data centres using an increasingly stable electricity supply, the class explores how a reliable internet connection is made possible in Singapore. With regard to the notion of multidisciplinarity, students are reminded that they, unlike their parents' generation, are now living and studying amid the Fourth Industrial Revolution, i.e., they can no longer stay within their comfort zone of their chosen course of study (e.g., law, accountancy, business, IT, and social science). A mini case study of driverless cars helps drive home the point that such a project (minimally) involves a good understanding of hardware, software, sensors, law, ethics, insurance, and government policy.

During the fifth week, students can organise their own fieldtrip to tackle project #1 (one of two projects students must complete) the results of which have to be presented in week 7.

The sixth session is devoted to a better understanding of some of the economic enablers of Singapore such as the role of leadership, governance, and innovation as well as the importance of creating and maintaining sustainable knowledge clusters (Evers *et al.*, 2018) and knowledge hubs (case examples include the city's offshore marine cluster and the WaterHub).

To nudge students into appreciating the importance of the 3Rs (reduce, reuse, and recycle), the seventh session features an excursion to one of Singapore's incinerators, such as the Senoko Waste-to-Energy Plant, which also helps them internalise some of the smart energy management imperatives (beyond the importance of 'good' energy-saving habits and renewable energy sources) around the 3 Rs. Another topic is the status of solar photovoltaic systems in Singapore and related opportunities for smart energy start-ups to disrupt the current structure of Singapore's electricity market. Other class topics include smart R&D Management (science and technology parks, R&D ecosystems, and case studies, e.g., Silicon Valley), smart healthcare (e.g., e-health implementation in local hospitals), and smart prisons (people, process, and technology).

During the eighth class meeting, students are introduced to the topic of 'Sustainable ('Smart') Mobility and Land Transport.' Key points include (i) how traffic jams cause more air pollution and (ii) how self-driving cars decrease accident rates due to the reduction of human errors. In-class role plays focusing on walkable cities and cycling cities are also conducted to help students better understand the roles played by the various stakeholders. Additional emphasis is put on teleworking and the role of ICT as enabler of smart mobility, categories of land transportation, policy issues related to public transport, driverless cars, road planning and design, road traffic management, and the social and legal implications of driverless car technology. The concept of smart cities is discussed in greater depth to encourage students to look beyond the attractiveness of innovative technologies so that they focus first on the core mission of city living; for example, it is more important that buses run on time before projects to Wi-Fi-enable bus journeys are initiated.

1.3. *Deliverables, Assessment Approach, and Project Topics*

To achieve the overall learning objectives, students are required to write a (i) term paper on a sustainability-related topic, (ii) organise a field trip to a sustainability-related site, and (iii) conduct a major student project to further internalise course-related knowledge and build competencies. Tables 1 and 2 illustrate the rubrics used for the first two assignments.

Table 1. Rubric for Term Paper

C grade	B grade	A grade
Partial research work done.	Sufficient research work done — using mainly internet sources.	Extensive research work done — beyond internet sources.
Subject matter is not fully understood. Scope is incomplete (*).	Reasonably clear understanding of the subject matter and scope.	Very clear understanding of the subject matter and scope.
Analysis is average.	Good analysis.	Excellent, thorough analysis.
Gaps in comments and recommendations.	Comments and recommendations are above average.	Insightful comments and recommendations.

Table 2. Rubric for Field Trip

C grade	B grade	A grade
Selected site is not relevant for achieving course objectives.	Selected site is relevant for achieving the course objectives.	Selected site is highly relevant for achieving the course objectives.
Research work is only partially done — it covers only basic background information of the site.	Research work is good and covers all the key features and some minor features of the site.	Excellent research work that covers all major and minor features of the site.
Presentation highlighted less than 75% of the features of the allocated site.	Presentation highlighted more than 80% of the features of the allocated site.	Presentation is insightful and highlighted almost 100% of the features of the allocated site.

Cumulative assessment (CA) constitutes 100% of the final grade, consisting of (i) Individual Assessment (60% of total) — in turn consisting of Class Participation (15%), Term Paper (20%), and MCQ Test (25%) — and (ii) Group Assessment (40% of total) — in turn consisting of a Minor Group Project/Field Trip (15%) and a Major Group Project (25%).

Components of Individual Assessments

Class Participation —Students are encouraged to ask questions and offer opinions in class. Active and well-thought-through discussions are expected from all students.

Term Paper — Each student will be assigned one smart city research topic and is required to produce a graded paper of between 2300 and 2500 words (not including the citations section).

MCQ Test — This comprises twenty-five multiple-choice questions testing students' understanding of all the smart city concepts covered in class, during site visits, and the assigned readings. The test will be held during Session 12.

Field Trip Project

The groups have to conduct their own research at the group level and can organise the field trip any time after Week 1. Each group will be allocated a place of interest in Singapore (as stated in Table 3), and students have to (i) prepare for the visit (pre-visit) by doing research on the site; (ii) visit the site (taking photos, studying posters/signboards if any, etc.); (iii) prepare for an in-class presentation (post-visit); and (iv) present their observations and findings in class.

Other than reading about the site, students are tasked to familiarise themselves with the UN Sustainability Goals (https://www.un.org/sustainabledevelopment/sustainable-development-goals/) and the EU Smart Cities framework: www.smart-cities.eu/model.html. They must also look for similar features in other smart cities around the world (e.g., by using the following non-exhaustive list for their research):

New York City	London	Paris	San Francisco	Sydney
Songdo	Boston	Hong Kong	Madrid	Shanghai
Beijing	Mumbai	Rio De Janeiro	Santander	Copenhagen
Barcelona	Kazan	Moscow	Innopolis	Bogota

Table 3.   Field Trip Locations and Focus Areas

	Location	Area of Focus
1	Biopolis or Fusionopolis *(choose one)*	R&D Hub
2	Hub at Jurong East	Building a town hub
3	Botanic Gardens	World heritage site
4	Esplanade Theatres	Cultural entertainment
5	Changi Airport	Aviation Hub, shopping, eating
6	NEWater Plant	Sustainable water supply
7	SMRT interchange (select any)	Smart mobility
	Alternative Sites:	
	Clarke Quay / Boat Quay (spare)	*Dining, nightlife*
	Punggol Waterway Park	*Community park*

Group Project #1 is due in Week 7. Each group must do a 15-minute presentation in class (no report is required).

1.3.1. *Major student project*

The major group project involves a deep(er) study into 'good' management practices with regard to the concept of digital sustainability at organisational levels to meet current and future challenges faced by various stakeholders, e.g., with regard to smart(er) energy usage, smart(er) mobility, and smart(er) consumption patterns. Each group will be assigned one major goal-oriented task and is expected to produce 'good' examples of sustainable practices and areas where an organisation is doing 'poorly' (e.g., no offsetting of carbon emissions). Students are expected to present their research-based project outcomes in a professional business format, aimed at making the value-added components of their (newly proposed) sustainability practice explicit to convince decision-makers to 'buy into' their idea(s). Each team must do a 15-minute presentation in class at the end of the course and submit a poster/info graphic, summarising the key components of the proposed solution. During the first run of the course, the following project topics were covered by students:

Suggestions for SMT Sustainable Digital City Projects (First Run of SMT112)

- Compute the digital carbon footprint of the SMT112 class over Term 1 and plant one (or more) tree(s) on campus as a compensatory carbon offsetting measure.
- Identify three truly sustainable cities and explain three best proven practices of offsetting a city's digital footprint.
- Develop a robust business canvas for a new clean tech start-up which can help make the Bras Basah district more carbon neutral.
- Measure the digital carbon footprint of SMU and suggest five new strategies to make the university (more) carbon neutral.
- Measure the outdoor or indoor air pollution with regard to fine dust affecting SMU and suggest ways to lower it drastically.
- Examine the waste-to-energy pilot project at Tampines Hub and discuss its transferability to SMU via a SWOT-inspired project plan.
- Study the carbon footprint annually of all the aircraft that take off and land at Changi Airport and make a recommendation to SIA about how to reduce the airline's carbon footprint.
- Utilise a relevant open dataset provided by the Singapore Government with regard to SMT112 course topics and create a blueprint (business canvas) for monetising these data.
- Develop an implementation plan to get 50% of SG households to recycle (more) by 2025.

Our experiences have shown that students are committed and engaged once they understand the wider context in which these projects are embedded. However, it also seems that the overall awareness of the sustainability agenda, especially when it comes to 'digital' sustainability, is still rather low compared with other countries (e.g., Germany). Therefore, more emphasis was put on the digital sustainability aspect during the second run of SMT112.

Suggestions for SMT Sustainable Digital City Projects (Second Run of SMT112)

- Discuss how SMU manages its carbon reduction efforts.
- Check out the website of UWCSEA and its overseas tree-planting project(s). Calculate the number of trees the class of SMT112 should plant to offset students' carbon emissions in Term 2 (focus on flights only).
- Discuss and provide some evidence: Are electric cars carbon neutral (yes/no)? If yes, explain why. If not, explain why not.
- Explore ISO14064-1:2006 and explain history, objectives, and outcomes.
- Study City Development Ltd.'s Integrated Sustainability Report (2016) and explain how the CDL manages its carbon reduction efforts.
- Calculate the carbon footprint of class SMT112 as far as students' mobile phone usage during Term 2 is concerned and suggest ways to reduce it.
- Assess the performance of SIA and Changi Airport as carbon emitters in comparison with other airlines and airports and suggest ways to reduce their emissions.
- Explore the European Carbon Trading system and check out how it works. What would it take to implement such a system in ASEAN (pros and cons)?

During both runs, several consultations with students took place to further fine-tune the scope of the projects. Table 4 presents the rubrics for these projects.

Table 4. Rubric for Main Group Project

C grade	B grade	A grade
The ideas presented (orally and on the poster) are uninteresting — either textbook ideas or copied from an existing idea. Implementation details are not well thought out, and the business plan canvas is incomplete. Upon further probing, students are unable to articulate potential challenges, e.g., how to obtain buy-in and funding. In its current form, the output would not be very useful in turning the sustainability ideas into practice.	The ideas (presented orally and on the poster) are somewhat interesting — attempts to think out of the box. Implementation details are quite well thought out, although some minor gaps exist in the business plan canvas. Upon further probing, students manage to explain some (but not all) of the most critical challenges associated with the proposal and how to tackle them. In its current form, the output could inspire further development of the sustainability practice despite some minor gaps.	The ideas (presented orally and on the poster) are very interesting and innovative, showing exceptional insight into the domain area. Very good implementation details based on a plausible, convincing (business) plan of action. Upon further probing, students manage to iron out potential stakeholder concerns regarding the implementation (e.g., lack of $) goals of the proposed urban sustainability solution (for a 'smarter' city). The project's output is clearly actionable and highly supportive of the UN Sustainability Goals.

2. Pedagogical Issues Observed and Solutions

One challenge for the instructors is to ensure that students appreciate and buy into the concept and importance of 'digital sustainability' in line with the course title. The first task of defining sustainability is easy: the term 'sustainable development' refers to the importance of meeting the needs of the present without compromising the ability of future generations to meet their own needs. Seen as the guiding principle for long-term global development, sustainable development consists of three pillars: economic development, social development, and environmental protection.

To clarify the term 'digital sustainability,' however, is more challenging. How do we do that? One approach is to link the concept of digital sustainability to the problem of increasing carbon emissions (which requires effective carbon emission management) and to explain proven ways to reduce one's carbon footprint once it has been calculated based on the example of sending emails and watching YouTube videos (both of which add to students' carbon footprint). Another focus is the air pollution problem and the need to come up with ways for 'cleaner air for smart cities.' We shall discuss these examples in the following in greater depth.

2.1. *The Carbon Footprint of Sending Emails*

Ever wondered how your emails may contribute to your personal carbon footprint? According to estimates published in Phys.org, sending a short email adds about 4 g of CO_2 equivalent (gCO_2e) to the atmosphere (an email with a long attachment has a 10-fold carbon footprint, i.e., 50 gCO_2e!). The carbon output of 65 emails is comparable to driving a mid-sized sedan passenger (petrol) car for about one kilometre. After five workdays, your 325 emails would have pumped out at least 1.3 kg of CO_2 into the atmosphere. Over a year, your emails would have contributed quite drastically to the global greenhouse effect and indirectly to global warming.

This is indeed 'An Inconvenient Truth,' as explained in the 2006 Oscar-winning environmental documentary film by Davis Guggenheim featuring former US Vice President Al Gore who triggered a global movement against the climate crisis. As citizens of a 'smart nation' undergoing digital transformation, and as consumers of digital products and services, are we knowledgeable about what is at stake?

According to Singapore's National Climate Change Secretariat, Singapore's 'business-as-usual emissions' are estimated to have amounted to 77.2 million tonnes (MT) in 2020 (Singapore is responsible for around 0.11% of global emissions). Climate change affects Singapore through rising annual mean temperatures — from 26.6°C in 1972 to

27.7°C in 2014. In addition, there has been an increased mean sea level in the Straits of Singapore at the rate of 1.2 mm to 1.7 mm per year between 1975 and 2009, and an uptrend in annual average rainfall from 2192 mm in 1980 to 2727 mm in 2014. Not surprisingly, residents have noticed increased episodes of flash floods, all exemplifying some of the effects of climate change on this island nation.

In 2019, the Ministry of Finance introduced a new carbon tax through the Carbon Pricing Act (CPA) to steer Singapore's transformation towards a low carbon economy. As stated on the website of the National Environment Agency (NEA), the carbon tax is applied to all industrial facilities with an annual direct GHG emissions of 25,000 tonne of carbon dioxide equivalent (tCO_2e). The initial carbon tax rate was set at \$5 per tonne for 2019 to 2023. The carbon tax rates will be raised to \$25 per tonne in 2024 and 2025, followed by \$45 per tonne in 2026 and 2027. By 2030, it will be reaching \$50 to \$80 per tonne. It will be interesting to see how effective the new policy will be in convincing private and public organisations that digital products are neither 'carbon light' nor 'low impact,' and that we are all accountable for the emissions. It remains to be seen whether the new carbon cost-related financial regulations or a formal ISO 14064 Carbon Emission Reduction Validation programme will effectively nudge stakeholders to adopt greener energy sources (Thaler & Sunstein, 2008; Vyse, n.d.).

An interesting challenge for educators is to create buy-in for the concept of digital sustainability among a generation of digital natives who contribute quite significantly to the problem of 'internet pollution.' Sending short messages, using photo- and video-sharing social networking services and searching the internet requires electrical energy which in turn causes a negative environmental impact. According to BBC's Science Focus, the internet is responsible for roughly one billion tonne of greenhouse gases a year, or around two per cent of world emissions. In 2017, Facebook (which has made a long-term commitment to be 100% renewably powered), with a total of 2.2 billion social network users, registered 979,000 metric tons of carbon dioxide equivalent (MT CO_2e).

What about streaming videos? Netflix itself also stated in 2014 that in delivering its service, its average customer had a carbon footprint of 300 g per year. It has since made its service carbon neutral; however, this does not factor in the power used by the devices when content is consumed. The carbon challenge of social media will no doubt further intensify with the rising number of users, which is expected to have exceeded 3 billion by 2021.

Powered by big data centres that house computing, server, and networking equipment, the rapidly increasing volume of digital transactions is arguably not sustainable given their large carbon footprint.

While there is a trend towards reducing the energy usage in data centres and evaluating their energy sources in favour of green(er) cloud-based hosting solutions, it requires more thinking to find ways to nudge individuals into concrete behavioural change to reduce carbon emissions as much as possible and offset as much carbon pollution as they emit.

One classroom approach is to use social norms to alter behaviour. Research has shown that informing people about their own energy use as compared to nearby residents (and how they can decrease energy consumption) can lead to significant long-term reduction in energy consumption to more sustainable levels. Imagine the behavioural and public relations impact if we begin to rank organisations by their carbon emissions.

Another educational strategy is to use a carbon footprint calculator to estimate one's own greenhouse gas emission level in comparison to one's peers. It would be better still if students can calculate emissions generated because of activities like watching YouTube, driving Mum's car to school, or using Scoot for a weekend trip to Bali. In addition, it is important that users know what it takes in terms of tree planting or financial contributions to effectively offset them.

Yet another tactic to obtain buy-in is to refer to the early (and visibly very successful) tree-planting initiative of Singapore's founding prime minister, the late Mr Lee Kuan Yew, which began with him planting a Mempat tree in Farrer Circus in 1963 that catalysed Singapore becoming a green 'City in a Garden.'

Any discussion about digital sustainability must go beyond meeting the digital needs of the contemporary generation. In order not to compromise the ability of future generations to meet their own needs, it is important to act now and to reduce the greenhouse effect which is warming Planet Earth. Everyone can start small by changing one's email habits by unsubscribing from all the financial market newsletters that are not read and skipping that non-essential internet search (0.2 gCO_2e on an energy-efficient laptop and 4.5 gCO_2e on an old desktop computer). One could also be bolder and offset one's personal carbon footprint by planting a tree. A Poplar tree, for example, can absorb about 300 kg of CO_2 and offset 1,000 air miles. As the proverb goes, 'The best time to plant a tree was 20 years ago. The second-best time is now.'

2.2. *Can AI Help to Cope with Air Pollution in 'Smart Cities'?*

Another pedagogical approach we use to sensitise students is to discuss the problem of air pollution and how it can be mitigated through technology. The prospects look promising for the deployment of artificial intelligence (AI) in the quest for a clean and liveable environment in urban settings. The use of deep technology and innovations such as AI is one area where Singapore can excel while also being the ideal test bed for such experimentation.

One of the salient points in the Singapore budget 2018 was the emphasis on a smart, green, and liveable city. In particular, the budget made specific reference to "emissions abatement." The government expects to collect carbon tax revenues of nearly S$1 billion in the first five years of its implementation, and is prepared to spend an amount more than the tax revenue collected "to support worthwhile projects which deliver the necessary abatement in emissions" (*The Straits Times*, 2018). Such fiscal nudges are timely because reducing emission intensity and volume can lower particulate air pollution and simultaneously create beneficial health risk reductions.

Particulate matter (PM), which consists of many hazardous substances, refers to the sum of all solid and liquid particles suspended in air. This includes organic and inorganic particles, such as dust, pollen, soot ('black carbon'), smoke, and liquid droplets. PM-induced pollution can result in chronic bronchitis or asthma and premature deaths because of cardiovascular problems or a stroke. PM2.5, for example, refers to inhalable pollutant particles measuring 2.5 microns and smaller in diameter — about a 30th the diameter of a human hair. These ultrafine particles are perilous because they can lead to a variety of health issues such as reduced lung function. Some may even be absorbed into the bloodstream. The economic cost of particulate air pollution on health in Singapore is substantial. An earlier study conducted by economists Euston Quah and Wai-Mun Chia estimated the total health costs associated with particulate matter in the air to be US\$3.75 billion or about 2.04 per cent of Singapore's GDP in 2009.

Vehicles represent a major source of particulate matter besides carbon dioxide and other tailpipe pollutants, such as carbon monoxide, hydrocarbons, and nitrogen oxides. According to data by the World Health Organization, 92 per cent of the world's population lives in places where air quality levels exceed WHO limits, resulting in millions of deaths every year. In 2012, 6.5 million deaths (or 11.6% of all global deaths) were associated with indoor and outdoor air pollution according to WHO estimates.

While more cities are now vigorously monitoring air pollution, comprehensive and actionable baseline data for monitoring progress in combatting it are not always available. Measurements are often done selectively based on a limited number of measurement stations. This makes it difficult to get reliable and valid environmental data, for example, about the sources of contaminants that could inform better policymaking for greener cities based on real 24/7 insights. Reducing harmful emissions and clean air predictions require precise, intelligent, and actionable measurement systems.

This is a pain point that can be addressed by harnessing the power of digital technologies such as sensors, Internet of things (IoT), and artificial intelligence. Through the computation of real-time IoT sensor data (with detailed spatial and temporal pollutant measurements) obtained from various sources such as ground-based sensor units and commercial satellites, user-friendly air quality heat maps and executive dashboards can be created (such as with the help of machine learning algorithms for predictive modelling) to determine the most severe pollution hotspots in order to take proactive steps towards further decarbonising the economy.

Examples of important use cases include tackling specific areas with higher concentrations of pollutants or leveraging (anonymous) crowd-sourced sensor data from cell phones to localise parts of the city which are harmful, with the aim of reducing one's own exposure to PM. Entrepreneurs keen to monetise air pollution data are well advised to do so on the basis of a digitalised business model that uses AI algorithms for creating green(er) and (more) liveable cities with better air quality via real-time assessment and management of outdoor air.

To stay ahead in the smart city race in an era of AI-enabled smart (urban) solutions, civil engineers, planners, regulators, and green start-ups keen to fight against air pollution should possess basic AI know-how and know-why over and above their expertise and experience in developing land use plans, revitalising urban precincts, making infrastructure services more accessible to disadvantaged segments of the population, or creating a scalable business model.

Smart city development programmes will have to be quickly rebooted with an AI advantage to leverage new digital technologies, big data analytics, and cloud computing for cleaner air and better public health. AI is arguably the new frontier of instrumented, environmentally friendly smart city initiatives.

AI-powered computer vision analytics with facial detection systems, in combination with environmental sensors mounted on smart lamp posts, for example, could enable authorities to identify in real time where exactly air pollution occurs, who is causing it, and where remedial and preventive action is required. As these systems can index faces to determine gender, race, and age, as well as perform facial matching against databases, rule violations may become outdated provided the ethics of such surveillance is feasible, guaranteed, and accepted.

Current technological advances in the area of AI-enabled smart city systems suggest that intelligent air quality management systems may soon become fully autonomous in making decisions, for example, remotely turning off the (combustion) engine of a self-driving car which is polluting the environment. The possibilities of smart, AI-enabled 'emissions abatements' are indeed endless. Singapore should seek a first-mover advantage through large-scale deployment. The demand for clean, green, and liveable cities is growing in tandem with the deep concern over climate change and how it intimately affects the health of people and the living environment.

Emission abatements will require companies to respond to this massive societal challenge in a purposeful manner. It is also the ethical and socially responsible thing to do with significant benefits beyond the health outcomes and economic well-being. Companies have to reduce their environmental footprint, with governments and consumers requiring that they do more to account for the negative externalities their business activities generate. AI-powered emissions abatement can be a powerful lever towards an empirically based targeted approach to dealing with environmental pollution and better public health. Singapore can and should exploit this window of opportunity to be a policy and thought leader in emissions abatement while spawning technological innovations to power the environmental drive.

3. Student Feedback and Learning Outcomes

SMU's evaluation approach centres on *reaction* evaluations. SMT112's performance (first run) was rated above average at around 6.2 on a 7-point scale. Students scored (very) well for projects but challenges were observed with regard to the quiz which was perceived to be demanding. The following are some typical responses by students when asked how they rate the course and what could be done better:

"The excursions involved in this course enable me to visit various interesting places which are related to the curriculum of this course. By visiting these places, I get to learn various methods on how a sustainable 'smart' city can be built and what measures can be taken to help in protecting the environment to build a sustainable environment."

"More activities can be conducted to help students in understanding and remembering every aspect of every excursion and relate them with the idea of sustainability."

"For the quizzes, give us past sem papers (I understand that they did not this time because this is a new mod but there could have been a sample) so that we have an idea what it would be like."

"The course should have a defined scope as it is very open ended ... There is indeed a sizeable amount of content under the broad umbrella of digital sustainable cities but there needs to be better focus. Clear instructions on projects should also be given and not changed halfway. Should have more congruence between textbook and slides."

These are valuable comments indeed, most of which were carefully considered before the roll-out of the second run. Besides the fact that the students did NOT make any comments about their main projects (which turned out to be very impressive indeed) and what they learned from those, they also point to the necessity of streamlining the core learning outcomes.

What then should the core goal-oriented learning outcomes of SMT112 be? Based on the experiences so far, we suggest the following:

- Appreciate that the development of a smart, climate change-friendly 'smart' city requires a multidisciplinary approach, i.e., the effective integration of technology (e.g., IoT skills), social science knowledge (e.g., observing urban spaces and conducting surveys), and management skills (e.g., project management).
- Know that impactful smart city solutions aimed at reducing our carbon footprint require interdisciplinary awareness and competencies (e.g., knowing the impact of rising CO_2 levels on environmental challenges such as urban warming) and the integration of IT in social, environmental, and urban policy contexts

(e.g., data analytics to translate the results of urban air pollution monitoring into concrete policy action).

- Draw connections between the various course topics and understand that the effective management of 'digital sustainability' (towards greater sustainability) requires effective 'nudging' so that people take concrete action to reduce their impact on climate change. "Nudge refers to a concept in behavioral science, political theory and behavioral economics which proposes positive reinforcement and indirect suggestions as ways to influence the behavior and decision making of groups or individuals" (Wikipedia).
- Evaluate the socio-economic benefits of ICT-enabled smart city solutions such as an IoT system to monitor air pollution and intelligent filtration systems as part of critical digital sustainability analysis skills.
- Create a prototype of a digital sustainability device which can help make a city 'smarter' and more sustainable, e.g., aimed at curbing the negative impact of increasing digital pollution.

The SMT 112 course experience underlines the importance of integrating insights gained in behavioural economics into the curriculum as suggested by research conducted by Hurlstone *et al.* (2014) and van der Linden *et al.* (2015) to enable concrete action by students.

4. Minor Course Syllabus Revamp

The SMT112 course underwent a minor revamp in August 2019. The field trip session was replaced by an additional ICT project that focused on real-world digital sustainability applications as shown in Table 5.

Learning-related questions students are tasked to explore via the various projects (as listed in Table 5) include the following: "If I am the project manager of such a turnkey project in the future after I have graduated, what are the important things I should look out for? How should I develop these skills now while

Table 5. New Projects in Lieu of Field Trip

Find an open-air car park near my location.	Find my car application (at Changi Airport).
Implement a solar farm on an area the size of a soccer field.	Smart lamp post project (in Country Utopia).
Use facial biometrics to find airport departure information.	Measurement of noise pollution (in a town centre).
Measurement of air pollution (near SMU location).	e-Waste recycling project (for a HDB estate).
Digital Advertising Signboards at 100 Bus Stops.	Mobile phone app to find free study spaces in SMU.

Table 6. Highlights of 3 Selected Projects

Implement a solar farm (in a polytechnic) on an area the size of soccer field	Figure out the senior management's objective for this project — is it a proof of concept or Phase 1 of a bigger solar energy project? Calculate the electricity output of the solar panels that can fit into a soccer field and how many buildings that output can power; make a post-implementation maintenance plan for the solar panels; and identify and mitigate the project risks.
Measurement of air pollution (near SMU location)	Figure out the scope and purpose of the air quality measurement, frequency of measurement, type of sensors and other equipment needed, which building rooftops to mount the sensors on, what kind of reports need to be generated and how frequently, and the post-implementation maintenance plan of the equipment; identify and mitigate the project risks.
e-Waste recycling project (for an HDB estate)	Figure out the main sponsor (stakeholder) of this project, calculate how many recycling bins are needed for the entire HDB estate, develop a communication plan to educate residents on the items that are accepted for recycling (e.g., other than laptops and mobile phones, can residents recycle old TV sets?), which external vendor to work with to remove the e-Waste items from the bins on a periodic basis, and the revenue model to be adopted; identify and mitigate the project risks.

I am still an undergraduate?" For each of the projects, students are required to elicit the user requirements; to determine the core infrastructure needed to roll out the project; to identify external stakeholders they need to engage; and to assess project-related risk areas. Table 6 exemplifies these challenges with regard to three project assignments.

By working on these digital sustainability projects, students are given a taste of real-world project management based on the 27 years of extensive project management experiences of one of the instructors who plays the role of a key stakeholder involved in each of the projects.

The role play included mimicking the typical answers given by key stakeholders of such projects where sometimes only partial answers are given due to busy stakeholders and/or stakeholders who may be new to the technologies and are therefore unable to provide more holistic answers to the project managers; this is good training for the students so that they will not be surprised in the future when they need to work on similar projects.

Part of the revamp also included a non-examinable talk on "How to ace a job interview" which the students found useful because one-third to half of each class comprised students in their final year.

The students found the new course format to be less content heavy in terms of covering academic (theoretical) aspects of sustainable digital cities and more aligned to helping them find related jobs and training in what to do when they are assigned digital city projects in their future workplace.

5. Conclusion

In this paper, we reflected on experiences in developing and teaching a new smart city-related course with a focus on the concept of 'digital sustainability' (in urban smart city contexts). We described the course context and pedagogical challenges, explained the significance of key teaching and learning contents and delivery

approaches (aimed at enabling students to appreciate the importance of managing digital sustainability), discussed related issues,
presented some of the projects conducted by students to internalise
the required content, and examined the conclusions with a view
towards streamlining key learning outcomes in relation to digital
sustainability challenges.

The overall course experience helped to further define the
somewhat nebulous concept of 'digital sustainability' beyond
existing definitions which put emphasis on sustainability or
sustainable development. In the context of SMT112, digital
sustainability is also about the need to reduce digital pollution,
e.g., the carbon footprint of electronic gadgets powered by
conventional fossil fuels so that the ability of future generations to
meet their own needs is not compromised. In the context of long-
term global development, this definition cannot be disconnected
from economic development, social development, and
environmental protection.

Student feedback suggests that site visits can play a key role in
helping learners appreciate the importance of managing digital
sustainability with a view towards minimising energy inputs in support of renewable energy sources. However, sensitising students so
that they might act to lower carbon footprints and/or to do something against climate change-induced air pollution is a very different
matter.

Van der Linden *et al.* (2015) have defined five best practices for
promoting (public) engagement with the critical issue of climate
change:

- 'Make it real' by increasing people's sense of urgency about climate change, e.g., let people experience it. While this is not
  possible in the SMT112 course, there are alternatives such as
  excursions to local sites where the impact of climate change can
  be felt. Besides an external visit, this might also entail a visit to a
  'hot' data centre (as practiced by us during one of the runs of
  SMT112).

- 'Use the herd,' e.g., by highlighting international polls showing strong majorities concerned about global warming who believe that climate change was caused by human activity. A localised example would be to show SMT112 students the current design of SP's (Singapore Power Limited) utility bills, which helps customers compare their own energy consumption with (the average of) their neighbours. According to nudging research, people tend to adjust their own use to conform to group norms.
- 'Make it more immediate' by taking note of local (and regional) effects of climate change rather than distant places. For example, one could be to organise a tour along Changi's coastal road to observe the (projected) effects of global warming on sea level changes in that area.
- 'Frame it differently' (consistent with the principle of loss aversion) so that people are not reluctant to pay more, e.g., for cleaner energy. One approach could be to frame necessary financial contributions (e.g., carbon tax) as reductions in future raises — rather than losses.
- 'Make it a feel-good thing' by emphasising the intrinsic benefits of environmental action (rather than providing extrinsic benefits). Research has shown that emphasising "the intrinsic, feel-good rewards of saving the planet is a more productive and sustainable strategy than using external rewards." In concrete terms, this might include jointly conducting a tree-planting activity.

References

Evers, H.-D., Gerke, S. & Menkhoff, T. (2018). Knowledge cluster development through connectivity: Examples from Southeast Asia. In T. Menkhoff, S. N. Kan, H.-D. Evers & C. Yue Wah (Eds.), *Living in Smart Cities: Innovation and Sustainability*. Singapore: World Scientific Publishing (Chapter 8).

Foo, S. L. & Pan, G. (2018). Singapore's vision of a smart nation — Thinking big, starting small and scaling fast. In T. Menkhoff, S. N. Kan,

H.-D. Evers & C. Y. Wah (Eds.), *Living in Smart Cities: Innovation and Sustainability*. Singapore: World Scientific Publishing (Chapter 1).

Hurlstone, M. J., Lewandowsky, S., Newell, B. R. & Sewell, B. (2014). The effect of framing and normative messages in building support for climate policies. *PLoS ONE*, 9(12), e114335. doi:10.1371/journal. pone.0114335.

Keivani, R. (2009). A review of the main challenges to urban sustainability. *International Journal of Urban Sustainable Development*, 1(1–2), 5–16.

Komninos, N. (2009). Intelligent cities: Towards interactive and global innovation environments. *International Journal of Innovation and Regional Development*, 1(4), 337–355.

Kong, L. (2018). Making sustainable creative/cultural space in Shanghai and Singapore. In T. Menkhoff, S. N. Kan, H.-D. Evers & C. Y. Wah (Eds.), *Living in Smart Cities: Innovation and Sustainability*. Singapore: World Scientific Publishing (Chapter 4).

Menkhoff, T., *et al.* (2018). What makes a city "smart"? In T. Menkhoff, S. N. Kan, H.-D. Evers & C. Y. Wah (Eds.), *Living in Smart Cities: Innovation and Sustainability* (pp. 1–59). Singapore: World Scientific Publishing.

Miller, S. (2018). Country 2.0 — Upgrading cities with smart technologies. In T. Menkhoff, S. N. Kan, H.-D. Evers & C. Y. Wah (Eds.), *Living in Smart Cities: Innovation and Sustainability*. Singapore: World Scientific Publishing (Chapter 3).

Paskaleva, K. (2009). Enabling the smart city: The progress of e-city governance in Europe. *International Journal of Innovation and Regional Development*, 1(4), 405–422.

Phang, S.-Y. (2018). Alleviating urban traffic congestion in smart cities. In T. Menkhoff, S. N. Kan, H.-D. Evers & C. Y. Wah (Eds.), *Living in Smart Cities: Innovation and Sustainability*. Singapore: World Scientific Publishing (Chapter 17).

Sassen, S. (2005). The global city: Introducing a concept. *Brown Journal of World Affairs*, 11(2), 27–43.

Thaler, R. H. & Sunstein, C. R. (2008). *Nudge: Improving Decisions about Health, Wealth, and Happiness*. New Haven, CT: Yale University Press.

The Straits Times. (2018, February 19). Singapore budget 2018: Carbon tax of $5 per tonne of greenhouse gas emissions to be levied. https://www.straitstimes.com/singapore/singapore-budget-2018-carbon-tax-of-5-per-tonne-of-greenhouse-gas-emissions-to-be-levied.

Valdez, A., *et al.* (2018, February). Roadmaps to Utopia: Tales of the smart city. *Urban Studies*, 5(7), 1–18.

van der Linden, S. (2015). Intrinsic motivation and pro-environmental behaviour. *Nature Climate Change,* 5(7), 612–613.

van der Linden, S., Maibach, E., & Leiserowitz, A. (2015). Improving public engagement with climate change — Five "best practice" insights from psychological science. *Perspectives on Psychological Science,* 10(6), 758–763.

Vyse, S. (n.d.). Behavior & belief — Nudging people to save the planet. https://skepticalinquirer.org/blog/nudging_people_to_save_the_planet/?/specialarticles/show/nudging_people_to_save_the_planet.

Wang, P. K. & Kiat, L. W. (2018). Towards a smart nation — It's about people, ultimately. In T. Menkhoff, S. N. Kan, H.-D. Evers & C. Y. Wah (Eds.), *Living in Smart Cities: Innovation and Sustainability.* Singapore: World Scientific Publishing (Chapter 2).

Williams, K. (2009). Sustainable cities: Research and practice challenges. *International Journal of Urban Sustainable Development,* 1(1–2), 128–132.

Chapter 7

City Developments Limited — A Business Case on Building Smart and Sustainable Cities

Esther An

1. The Rise of Sustainability in the New Economy

1.1. *Convergence of Global, National, and Business Commitment*

The world is in a climate emergency. The warmest period in the past 170 years has been the last seven years, making the past 10 years from 2012 to 2021 the warmest decade on record.[1] In 2021, the Intergovernmental Panel on Climate Change's 6th Assessment Report was deemed a 'code red for humanity.' The Working Group II's contribution to the 6th IPCC Report declared that there is a 'rapidly narrowing window of opportunity to enable climate resilient development,'[2] and warned of severe impacts if the world exceeds global warming of 1.5°C.[3]

---

[1] 2021 Among Earth's Hottest Years, UN Says as Climate Meetings Start. Bloomberg. 31 October 2021.

[2] IPCC_AR6_WGII_FinalDraft_FullReport.pdf.

[3] Press release | Climate Change 2022: Impacts, Adaptation and Vulnerability (ipcc.ch).

151

In today's world, embracing sustainability is no longer a choice. In the global Race to Zero[4] led by the UNFCCC, over 7,197 businesses, 1,103 cities, and 541 investors have stepped up their ambition and joined the global alliance to catalyse climate change. Following COP26, an impressive 90% of global GDP has committed to achieving net zero by or near mid-century.[5] According to the 17th Edition of the World Economic Forum Global Risks Report, environmental risks were perceived to be the five most critical long-term threats over the next 10 years.[6] Climate risks are investment and business risks — the damage caused by climate change is projected to result in an increase of up to 41% of global property premiums until 2040.[7]

At the national level, the Singapore Green Plan 2030 was launched in February 2021, bringing sustainability to the forefront of the national agenda. In February 2022, the Singapore government announced an enhanced climate ambition and commitment for the country to achieve net-zero emissions by or around 2050. The Singapore Budget 2022 also included plans to progressively increase carbon tax from 2024 to 2030, in line with plans to achieve the national net-zero goal.

The business case for Environmental, Social, and Corporate Governance (ESG) integration is stronger than ever today. We have seen sharp spikes in companies pledging support towards the global agenda for a net-zero future. It is no longer about short-term profitability but acknowledging that sustainability is material to a company's long-term growth. By integrating sustainability into core business strategies, companies retain their competitive advantage and build resilience. Embracing the triple-bottom-line approach pushes companies to do well, while doing good. Only when

---

[4] Home — Climate Champions (unfccc.int), March 2022.

[5] COP26 signals accelerated zero carbon investment drive; severe climate risks remain — Investor Group on Climate Change (igcc.org.au).

[6] WEF_The_Global_Risks_Report_2022.pdf (weforum.org).

[7] In a world of growing risk the insurance industry has a crucial role to play | Swiss Re.

businesses start to see sustainability in a new light will they be able to spur innovation, create value, and turn green into gold.

CDL's Environmental, Social, and Governance (ESG) strategy and its firm commitment to 'Conserving as We Construct,' the corporate ethos established in 1995, has positioned it well in the transition to a low-carbon economy. Its value creation business model, anchored on four key pillars — **Integration, Innovation, Investment, and Impact** — provides CDL a solid foundation to mitigate and adapt to unprecedented threats and challenges. CDL is committed to achieving three deliverables: 'Decarbonisation,' 'Disclosure and Communication,' and 'Digitalisation and Innovation.'

2. Integrating Sustainability into CDL's Strategy and Operations

2.1. *Longstanding and Unwavering Leadership Commitment*

Integrating sustainability at the highest governance level in CDL enables strategic oversight of ESG issues for long-term value creation. Since 2017, the leadership spearheaded by CDL's Board Sustainability Committee has been critical in delivering CDL's sustainability purpose, integration, and performance that add business and economic value to the company.

CDL has established the longest history of having a dedicated sustainability portfolio amongst Singapore companies, headed by the Chief Sustainability Officer (CSO) since 2014. Consolidating the CSR and environmental conservation strategies and initiatives started in 1995, CDL's sustainability portfolio was officially established to lead sustainability in a dedicated department, reporting directly to the Board Sustainability Committee. Its role is to spearhead and implement ESG integration into CDL's business, operations, and growth strategy. In 2018, the CDL Group introduced its **G.E.T.** strategy — focusing on **G**rowth while adopting an ESG lens, **E**nhancement of assets to drive operational efficiency, and **T**ransformation to deliver long-term and sustained value.

Figure 1.   CDL's Sustainability Governance Structure

2.2. *Integrated and Future-fit ESG Strategy: CDL Future Value 2030 Sustainability Blueprint*

Sustainability integration is critical in CDL's strategy to create long-term value and future-proof its business. In the face of global climate, health, and economic challenges, CDL's early adoption of ESG integration has enabled it to remain steadfast as a responsible business. The **CDL Future Value 2030** Sustainability Blueprint, implemented in 2017, maps out its strategic goals and ESG targets, and has remained effectively integrated into CDL's business strategies and operations. Since 2017, CDL has continued to track and report its performance annually from 2008 and every quarter.

CDL'S value creation model has adopted the concept of six distinct but interrelated capitals of the International Integrated Reporting Council (IIRC): financial, manufactured, organisational, human, social and relationship, and natural.[8] This expands CDL's value creation to consider more than just the financial aspects — tangible assets and liabilities — but to also consider continuing access to essential natural resources.

---

[8] IR-Public-Sector.pdf (cimaglobal.com).

2.3. *Integrated Sustainability Reporting and ESG Disclosure: What Gets Measured Gets Managed*

With climate change becoming a salient issue of our time, it is inarguable that climate risks have become investment risks. Sustainability reporting is therefore a critical step for businesses to take that helps them set goals, measure performance, and manage sustainability-related impacts and risks. Being the first Singapore company to publish a dedicated sustainability report in 2008, CDL has benefited from the process of producing 15 sustainability reports. What gets measured gets managed is very true. Target setting, tracking, and reporting have helped CDL identify gaps, take strategic action to improve and future-proof its business, and raise operational performance.

Over the years, CDL's sustainability reporting has evolved into a unique blended model, harmonising various international reporting frameworks, standards, and approaches. These include the GRI Standards as CDL's core since 2008. To address the diverse expectations of stakeholders, CDL has embraced the CDP since 2010, the Global Real Estate Sustainability Benchmark ('GRESB') since 2013, the Integrated Reporting Framework of the Value Reporting Foundation ('VRF') since 2015, Sustainable Development Goal ('SDG') Reporting since 2016, the framework of the Task Force on Climate-related Financial Disclosures ('TCFD') since 2017, and standards of the Sustainability Accounting Standards Board ('SASB') for Real Estate Sector since 2020.

Since 2014, CDL started conducting materiality assessments annually to determine the key economic, environmental, social, and governance (EESG) issues that are important to CDL's stakeholders. Due to COVID-19, CDL conducted more comprehensive materiality assessments both in 2020 and 2021, to be in closer alignment with the shifting priorities and expectations of its stakeholders. Today, the top five issues are climate resilience, energy efficiency and adoption of renewables, innovation, stakeholder impact and partnerships, and product/service quality and responsibility, reflecting the climate emergency we face. This annual exercise has helped CDL align the evolving expectations and deliverables amongst internal and external stakeholders and stay focused on issues that matter the most to them and CDL's business.

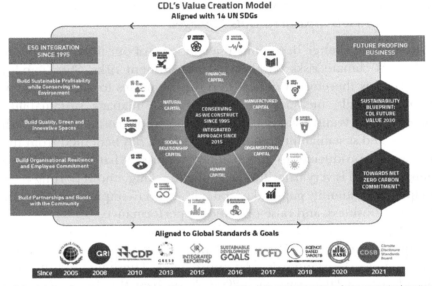

Figure 2. CDL's Value Creation Model

COP26 has resulted in a push for 'game-changing' policies and a synergy of multilateral collaboration to limit global warming to 1.5°C. It is therefore no surprise that climate resilience and energy efficiency have emerged as the top two material topics in 2021. This bodes well for CDL's longstanding dedication to green and healthy buildings.

2.4. *Embracing Supply Chain for Sustainable Design, Procurement, Construction, and Management*

In a constantly shifting environment fraught with challenges presented by the pandemic and climate change, building resilience into the supply chain is key. Supply chain risk management and robust systems and processes are essential to prepare any business for disruptions that threaten to weaken supply chain capability.

In adopting a forward-looking approach to tackle climate-related risks, CDL conducted a Supply Chain Segmentation Study

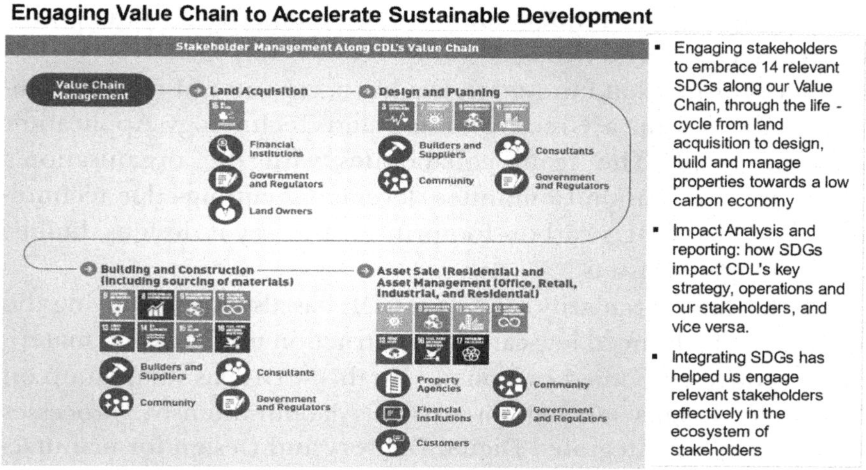

Engaging Value Chain to Accelerate Sustainable Development

- Engaging stakeholders to embrace 14 relevant SDGs along our Value Chain, through the life - cycle from land acquisition to design, build and manage properties towards a low carbon economy

- Impact Analysis and reporting: how SDGs impact CDL's key strategy, operations and our stakeholders, and vice versa.

- Integrating SDGs has helped us engage relevant stakeholders effectively in the ecosystem of stakeholders

Figure 3.   Stakeholder Engagement Along CDL's Value Chain

to identify CDL's sourcing and supply chain risks. The study was conducted on CDL's top 100 suppliers and top five building materials in 2019 and completed in 2020. The study helped to strengthen CDL's understanding of potential risk hotspots within the supply chain and improved its supply chain strategy.

For CDL's efforts to effectively engage its supplier network to address climate change, it was the only real estate company in Southeast Asia and the only Singapore company to be recognised as a 2021 CDP Supplier Engagement Leader for the second consecutive year. This places CDL amongst the top 8% of companies assessed by CDP for supplier engagement on climate change.

3. Innovation is Key to Green the Built Environment

3.1. *Digitalisation and Sustainable Technologies for a Green Revolution*

A low-carbon future is not possible without smart and sustainable solutions. With innovation identified as CDL's top ESG issue from 2017 to 2019, the company has intensified its search and application of viable green technology solutions, while tapping into the power of cross-sector partnerships.

Recognising that innovation is a key accelerator of climate solutions, CDL set up a Green Building and Technology Application team in 2020. The team collaborates with the organisation's Enterprise Innovation Committee, leveraging cutting-edge technology to reduce CDL's carbon footprint in the way it designs, builds, and manages its assets.

To advance circularity solutions, CDL has also been studying the feasibility of advanced low-carbon construction methods and materials to reduce embodied carbon. To do this, CDL has ramped up on buildable designs moving towards less labour-intensive processes and focuses on Integrated Digital Delivery and Design for manufacturing and Assembly (DfMA) technologies. Through this, CDL can reduce reliance on on-site workers, enhance workplace safety and health, and drive productivity improvements in construction and facility management.

CDL's strategic R&D partnership with the NUS School of Design and Environment since 2017 has served it well. In 2020, the NUS–CDL Smart Green Home developed an Acoustic Friendly Ventilation Window prototype that reduces noise while achieving air change efficiency of up to four times more than conventional windows.

Accelerating the shift towards renewable energy, CDL partnered with the Solar Energy Research Institute of Singapore (SERIS) to apply for a Solar Competitive Research Programme in 2020, to test-bed cost-effective high-power-density Building-Integrated Photovoltaics (BIPV) modules with the potential for implementation at CDL's properties. The company also piloted printed bifacial BIPV panels at City Square Mall as a test bed for more future installations.

Complementing CDL's target to achieve net zero for its buildings by 2030, CDL established a new Smart, Sustainable, and Super-Low-Energy (3S) Green Building Framework. This is an expansion of CDL's green building and green procurement guidelines, representing a holistic framework aligned with the BCA SLE building requirements, as well as international standards for

CDL and SERIS piloted a new generation of PV art wall (bifacial BIPV panels with prints) at the Singapore Sustainability Academy at City Square Mall in 2020. This serves as a test bed for more efficient PV installations.

Figure 4. PV Art Wall (bifacial BIPV panels with prints) at CDL's Singapore Sustainability Academy, City Square Mall

advancing health and well-being in buildings. In 2021, CDL updated the 3S Green Building Framework to include embodied carbon management in alignment with the Net Zero Carbon Buildings Commitment of the expanded World Green Building Council (WGBC). CDL is accelerating its innovation and digitalisation efforts to scale up retrofitting of existing assets to be certified as BCA Green Mark Super Low Energy (SLE) compliant. CDL will continue to engage its tenants and supply chain to adopt energy-saving fittings and sustainability measures, through the CDL Green Lease Partnership Programme and related business guidelines to achieve a low-carbon future.

CDL also has a finger on the pulse of the latest innovation trends and is an active investor in PropTech venture capital funds, such as Fifth Wall, Taronga, and Dragonrise Capital. CDL also actively explores and develops PropTech solutions in-house, such as digi-HUB. A digital platform developed by CBM Pte Ltd., CDL's whole-owned subsidiary, and a leading facility management company in Singapore, digiHUB focuses on predictive and integrated facility management solutions.

The SSA is the first ground-up initiative and zero-energy facility in Singapore dedicated to capacity building and thought leadership for climate action. Since its opening in 2017, it has tapped on 3,200 sq ft of solar panels on its rooftop as its energy source. The entire facility is built with over 80% of structural materials that come from sustainable sources.

Figure 5. CDL's Singapore Sustainability Academy

To achieve a net-zero world, zero-energy buildings are the way forward. To date, CDL has built two net-zero facilities using eco-friendly technologies — the Singapore Sustainability Academy (SSA) and the CDL Green Gallery at the Singapore Botanic Gardens. The SSA, a BCA Green Mark Platinum-certified building, is the first in Singapore to have its construction materials, Cross Laminated Timber and Glued Laminated Timber, verified by the Nature's Barcode™ system as coming from responsible sources.

3.2. *Decarbonising towards a Net-Zero Future*

In February 2022, the Singapore government set an enhanced ambition for the country to achieve net-zero emissions by or around 2050. The Singapore Budget 2022 also laid out plans to progressively increase carbon tax from 2024 to 2030, to help achieve the national net-zero goal. According to the World Bank Carbon Pricing Dashboard, the bold hikes put Singapore ahead of its regional counterparts such as Japan, China, and South Korea while lagging behind European pacesetters such as Switzerland and Sweden for carbon taxes per tonne of emissions.[9]

---

[9]What to make of the carbon tax increase in Singapore? — TODAY (todayonline. com), 7 March 2022.

Carbon tax is a key price signal and a powerful driver for businesses and individuals to lower their carbon footprint. The upcoming implementation of the tax hike in 2024 therefore presents corporates with an opportunity to strategically identify climate-related risks and explore green growth solutions.

CDL has been an early adopter of decarbonising its operations for over a decade now. CDL has constantly raised the bar for decarbonisation targets with robust tracking, monitoring, and evaluation to ensure its existing green building initiatives can mitigate rising operation costs that will come with the progressive carbon tax.

With the global race to zero fast gaining traction, CDL takes pride in supporting global and national climate goals and has pledged to commitments aligned with the COP26 Climate Pact. In February 2021, CDL became the first real estate conglomerate in Southeast Asia to sign the Net Zero Carbon Buildings Commitment of the World Green Building Council (WorldGBC). This is a global pledge to achieve net-zero operational carbon by 2030, covering new developments and existing wholly owned assets under CDL's direct management and operational control. In November 2021, during COP26, CDL joined 44 pioneering companies worldwide to extend its pledge towards a net-zero whole-life carbon emissions approach. Through this expanded commitment, CDL pledged to achieve maximum reduction of embodied carbon in new developments, compensating for any remaining residual operational and upfront embodied emissions via offsetting for new developments by 2030 and for all buildings to be net-zero carbon by 2050.

CDL also stepped up its decarbonisation commitment, aligning with even more ambitious carbon emissions reduction targets that have been assessed and validated by SBTi. The revised targets will help accelerate climate action to limit global warming to 1.5°C and support CDL's WorldGBC Net Zero Carbon Buildings Commitment.

Since 2009, CDL has started adopting carbon neutrality to supplement its efforts to search for innovative solutions to raise energy efficiency and transition to renewable energy to reduce its carbon footprint. Through carbon offsetting, CDL has achieved carbon neutrality for its headquarters' operations for more than a decade. In November 2021, CDL was the only Singapore real estate company amongst 19 pioneering companies invited to Climate Impact X's pilot auction. Climate Impact X is a partnership by SGX, DBS, Standard Chartered, and Temasek to provide a global exchange and marketplace for companies to access high-quality carbon credits. Through the auction, CDL successfully secured carbon credits from eight global natural climate solutions projects. These credits will be retired over the next three years to offset an estimated 6 to 7% of emissions from CDL's operations per year to complement its net-zero targets by 2030. CDL will continue to explore more of such innovative solutions to support the decarbonisation efforts.

While CDL has made good progress over the past two decades, there is always room to do more. CDL will continue to embrace change, transform its business, and champion innovative solutions to future-proof its business and create value for its stakeholders. Racing to net zero will remain the cornerstone for CDL's sustainability strategy in this decade of action.

4. Investment — Financing the Transition to a Green Environment

As the demand for green financing grows in the acceleration towards climate action, companies with strong ESG performance will gain better access to the fast-growing ESG investment funds and sustainable finance. Sustainable bond issuance set a new height globally in 2021, surpassing US$1 trillion for the first time ever, a 45% increase over 2020.

With sustainable finance emerging as a powerful enabler in building back greener and better, the urgent need to mitigate and adapt to climate risks is opening huge investment opportunities. This is evident from the UN Principles for Responsible Investment (PRI), an investor initiative in partnership with the UNEP Finance Initiative and the UN Global Compact, of which CDL became a signatory in September 2021. As of November 2021, the total Assets Under Management (AUM) of companies that are committed to PRI was more than US$121 trillion, signalling a rapidly accelerating transition towards sustainable assets. From January 2022, CDL's CSO

CDL's Republic Plaza Green Bond was the first green bond issued by a Singapore company in April 2017.

Figure 6. CDL's Republic Plaza

was appointed as a member of PRI's inaugural Real Estate Advisory Committee, which succeeds the Property Working Group coordinated by the UNEP FI.

In December 2021, CDL rolled out its Sustainable Investment Principles (SIP). This reinforces the company's commitment to taking proactive action in assessing potential portfolio risks and opportunities for sustainable investment decisions. The SIP complements its extensive list of ESG policies and guidelines. Furthermore, the SIP is formulated in line with the Glasgow Climate Pact and aligned with the global best practices laid out in the United Nations Sustainable Development Goals (UN SDGs), UN PRI, TCFD, UN Environment Programme Finance Initiative (UNEP FI), and other relevant frameworks.

Since issuing its first green bond in 2017, CDL has amassed more than S$3 billion in sustainable finance, including various green loans, a green revolving credit facility, and a sustainability-linked loan. In April 2021, CDL's South Beach Consortium secured a 5-year green loan totalling S$1.22 billion — one of Singapore's largest green loans to date. In August 2021, CDL and its JV partner jointly secured green loans amounting to S$847 million for the financing of two upcoming developments in Singapore — Piccadilly Grand and Tengah Garden Walk EC. As a green developer, CDL is heartened that its strong sustainability track record enables the company to tap into the fast-growing sustainable financing pool to benefit its joint venture projects, sharing a green vision of a low-carbon future with like-minded partners.

For the successful R&D and pilot of DigiHUB, a digital platform to raise building management efficiency, CDL secured a discount on the SDG Innovation Loan provided by DBS Bank. This represents the first time a Singapore entity has achieved a discount on a sustainability-linked loan through the adoption of an innovative project that supports the UN SDGs on a large-scale basis.

5. Impact: Building Sustainable Communities to Accelerate Climate Action

5.1. *Business Impact: Global Recognition of CDL's ESG Performance*

CDL's ESG performance is widely recognised by leading global sustainability benchmarks. These include the Global 100 Most Sustainable Corporations in the World by Corporate Knights, where CDL ranked fifth overall and maintained its position as the world's most sustainable real estate company for the fourth consecutive year. Other notable accolades include maintaining a double 'A' in the 2021 CDP Global A List for corporate climate action and water security. This makes CDL the only company in Southeast Asia and Hong Kong to score 'A' for corporate climate action for four consecutive years and water security for three consecutive years. CDL has also maintained an 'AAA' leader rating by MSCI ESG Research since 2010.

In November last year, CDL was one of 45 global companies and the only Singapore company to be awarded the inaugural 2021 Terra Carta Seal by His Royal Highness the Prince of Wales, through his Sustainable Markets Initiative. Presented during COP26 at Glasgow, the Seal recognises global companies driving innovation and demonstrating commitment to creating genuinely sustainable markets.

To make sustainability count, measuring a company's ESG impact is crucial. Social impact efforts that create real value

Figure 7.   CDL's ESG Track Record

differentiate one company from another. When our planet flourishes, people, communities, and the economy will thrive. It is pivotal to engage, educate, and empower people and communities for a more climate-resilient future.

5.2. *Social Impact: Driving Change through Advocacy, Capacity Building, and Community Partnerships*

The Singapore Sustainability Academy (SSA) was designed and built by CDL as Singapore's first ground-up, zero-energy facility dedicated to advocacy, capacity building, and the SDGs. It involved an extensive partnership comprising six government agencies and 15 founding industry and non-governmental organisation partners. Since its opening on 5 June 2017 till the end of 2021, the SSA hosted more than 630 sustainability-related training programmes and advocacy events, attracting close to 22,300 attendees. It has become a hallmark of CDL's community engagement and Singapore's leading knowledge and networking hub for sustainability, providing industry training, networking, and capacity building for climate action and implementation of relevant SDGs. Due to COVID-19, the SSA went online, reaching out to thousands of participants from over 25 countries, including Switzerland, Nigeria, Malaysia, Thailand, and the US.

5.3. *Women's Empowerment and Developing Future Eco-Champions*

Recognising that female empowerment is crucial for climate action, CDL created Women4Green. This collaborative platform aims to empower female executives to adopt and champion sustainable living at work, home, and play, supporting SDG 5: Gender Equality and SDG 13: Climate Action. Since its inception in late 2017, various initiatives and events were organised, covering themes like sustainable diet, fashion, and jewellery.

Nurturing youths has always been CDL's priority. Initiated by CDL in partnership with the NLB in 2013, My Tree House is the world's first green library for children, created to encourage environmental literacy and appreciation amongst kids. In 2020 and 2021, CDL and the NLB jointly organised My Tree House's Eco Storytelling Contests, promoting family bonding through storytelling, and encouraging young eco-champions to develop a love for nature and to take climate action.

Opened in 2013, My Tree House is the world's first green library for children ages 3 to 12.

Figure 8. CDL's My Tree House

5.4. *Capacity Building for Sustainability Practitioners*

With global and national efforts towards a low-carbon economy accelerating, it is estimated that ESG investment will hit US$1 trillion by 2030.

Around the world, demand for ESG talent is fast growing across sectors. The need is exceptionally acute in Asia, with its rapid urbanisation and economic transformation. The competition for talent is magnified by the gap between supply and demand of skilled and experienced talent. In the new economy, jobs have changed. New jobs have emerged and replaced existing ones. In tandem with this, the related required skills and competences have also rapidly evolved. With the pledge to net zero by or around 2050, the Singapore government has stepped up its nationwide plan to steward change towards a low-carbon future and announced various measures to achieve this.

A key tenet of effecting change is equipping the workforce to contribute to a more sustainable future. As part of the efforts to empower businesses and build knowledge capacity, multiple programmes have been launched. One of these is the S$180 million Enterprise Sustainability Programme to support Singapore companies in their sustainability journeys. With Singapore's accelerated pursuit of sustainable development, 55,000 jobs are expected to be created in the sustainability sector in the next 10 years.[10]

Being an early adopter of sustainability integration and reporting, CDL has built up over two decades of know-how and experience to support Singapore and the building industry's transition to sustainable practices. In partnership with Global Green Connect, CDL launched the *Sustainability Connect*, a platform for sustainability practitioners by sustainability practitioners in January 2022. This collaboration expands CDL's focus on capacity building and aims to equip sustainability professionals with practical knowledge, allowing them to scale up their skillsets and support their employers to

---

[10] Addendum to President's Address: 55,000 new jobs to be created in sustainability sector in next 10 years, says Grace Fu | The Straits Times (27 August 2020).

future-proof their businesses. This collaboration will be backed by key industry partners including the United Nations Institute for Training and Research (UNITAR), World Green Building Council (WGBC), Singapore Green Building Council (SGBC), Asia Pacific Real Assets Association (APREA), Singapore Institute of Directors (SID), SkillFuture Singapore, Employment and Employability Institute (e2 i), and Global Compact Network Singapore (GCNS). The Singapore Sustainability Academy will host training and interaction activities organised by Sustainability Connect.

6. Looking Ahead: Zero in on Future Value for Planet and People

On the trajectory of existing commitments, carbon emissions are set to rise 13.7% by 2030. To limit warming to 1.5°C, decarbonisation efforts need to be redoubled this decade.[11]

The Race to Zero campaign serves as a heartening example of the largest alliance committed to net-zero carbon emissions by 2050. It includes stakeholders in more than 60 regions, 1,000 cities, and 5,000 businesses.[12] With 90% of the world's GDP now covered by net-zero commitments,[13] there is real collective power to drive climate action. However, the stakes are high, and pledges need to be accompanied by concrete action.

Today, the business of business is no longer just business. Committing to the Race to Zero will add purpose to the Triple Bottom Line. As Singapore's real estate pioneer and green building leader, CDL has aligned its business with global and national goals to mitigate the negative impact of climate change. With CDL's Future Value 2030 Sustainability Blueprint established in 2017 as the

[11] https://www.weforum.org/agenda/2022/01/climate-change-action-trends-2022-un/.

[12] https://racetozero.unfccc.int/.

[13] https://www.eco-business.com/press-releases/what-are-the-pathways-to-achieving-net-zero-cities-in-asia/.

bedrock for its ambitious ESG goals and strategies, it is well placed to accelerate its net-zero goals. Contributing to a sustainable future will remain the purpose of CDL's business. Climate threats on the planet have a severe impact on people and businesses, and no one can thrive without a healthy planet.

Chapter 8

The Handprint Approach

Simon J. D. Schillebeeckx and Ryan K. Merrill

1. Introduction

Handprint Tech is a Singaporean social enterprise building the digital infrastructure of the regenerative economy. Born with the ambition to embed positive impact into every product and service on earth, Handprint serves as a case study in digital sustainability.

Handprint's premise grew from a series of investigations into using new technology in the convergence ecosystem of AI/machine learning, IoT, blockchain, and remote sensing to solving global environmental problems. In 2017, Professors Simon JD Schillebeeckx and Gerry George began a deep dive study of innovation into the natural world within the Lee Kong Chian School of Business at Singapore Management University (SMU). At a symposium on managing climate resources at the Academy of Management Annual Meeting, Simon met another young academic, Ryan Merrill, who had just finished his PhD in Climate Policy at the University of Southern California and was seeking a research position from Bangkok. In striking range of Singapore with a 'can-do' attitude, Ryan had just the right profile and mindset to round out the team.

Over two years, Simon and Ryan wrote multiple case studies on entrepreneurial ventures engaged in environmental innovations. One of these explored *the story of Worldview International Foundation,* a Norwegian non-profit pioneering mangrove restoration project in Myanmar's Bay of Bengal with a million dollars in decentralised crowdfunding (ICO) via the Heyerdahl Climate Pioneers Tree Coin (Schillebeeckx & Merrill, 2018). Simon and Ryan spent a week in Ayeyarwaddy, studying the benefits of mangroves and were so inspired by the octogenarian Executive Secretary of Worldview, Arne Fjortoft, that they founded Global Mangrove Trust (GMT) less than a month later. GMT is a Singaporean non-profit building fintech solution for peer-to-peer crowdfunding of community-based mangrove forestry.

18 months later, the two young academics published their report on Sustainable Digital Finance in Asia (Merrill *et al.,* 2019). Commissioned by DBS Bank and the United Nations and informed by GMT's sometimes challenging work with DBS (Schillebeeckx & Merrill, 2021), the report exposed major problems with voluntary carbon markets' abilities to support regenerative field work across South and Southeast Asia. Firms seeking to claim to be 'Net Zero' or 'Carbon Neutral' buy carbon credits representing a ton of avoided or removed emissions to neutralise pollution that cannot yet be abated by cleaning up value chain operations. While acknowledging many established issues with carbon credits (non-scalable verification, questionable additionality, difficulty differentiating between credit types, etc.), the authors honed in on problems in the inherent value of carbon credits to fund actual impact.

The research shows that credit pricing remained incredibly opaque, and that up to 80% of the money spent on buying credits can be absorbed by intermediaries like auditors, resellers, verifiers, project developers, and more resellers. As a result, of $20 spent on a carbon credit, as little as $4 would reach the local communities implementing true restoration of carbon-absorbing forests. The inspiration

seemed clear, if one could remove or drastically reduce these interme-
diary costs, **one could have five times more carbon absorption** without
spending a single dollar more on carbon removals. And if the marginal cost of carbon removals could so drastically reduce, more firms
and families may very well be tempted to start investing in climate
action.

2. From Non-profit to For-profit

While Simon and Ryan understood these dynamics in theory, it took
a series of meetings with future Handprint CEO, Mathias Boissonot,
to reveal the business opportunity. An engineer and planet-hacker,
Mathias led a consultancy securing research and development
grants for tech firms. He had recently led a management buy-out of
his firm and taken it from a loss-making to a profitable business in
18 months, which gave him a lot of credibility with his future co-
founders. Mathias was intrigued by the work GMT was doing in
collaboration with DBS on financial transparency and blockchain,
and proposed to work with GMT pro bono to develop a grant application to accelerate the development of the blockchain technology
the team had been working on. The grant application was never
successful due to a variety of issues, but as Mathias, Ryan, and Simon
spent more and more time talking about the founding ideas and
ideals of Global Mangrove Trust, Mathias became convinced that
these ideas and the early-stage technology GMT had been developing with DBS were solid ground upon which to start a business. He
quit as managing director of his company and started working full
time on what is now Handprint Tech at the end of 2019. The first
developer was hired in March 2020.

By August 2022, Handprint had raised over 2M USD in funding
and was negotiating an additional 1M investment. The team had
grown to thirty people in nine countries with a single goal in mind:
How do we make positive impact creation valuable for business?
Handprint's official documentation has the following mission and
vision statements:

2.1. *Mission*

Handprint is on a mission to integrate positive impact into every business. We connect companies to causes they and their customers care about. On the one hand, we curate and monitor diverse impact projects within the regenerative economy, focusing on habitat preservation, mangrove reforestation, coral reef reconstruction, and ocean plastic clean-ups as well as various types of social impact projects aligned with the UN Sustainable Development Goals. On the other hand, we create (plug-and-play) digital technology solutions that empower our clients to integrate positive impact into their transactions, activities, products, and services. Our suite of solutions forms a modular digital regenerative toolbox from which clients can select components that automate, integrate, and visualise positive impact creation. As such, our clients' existing products and services become regenerative in a few clicks so that they can grow with the planet.

2.2. *Vision*

Our vision is to build the digital infrastructure for the regenerative economy (both web2 and web3). To this end, we collaborate with partners across industries to embed regenerative actions into transactions, activities, products, and services. This includes banking (Idemia), advertising (Teads, Dentsu), sports (ClassPass), remittances (Thunes, DTOne), customer engagement (Braze), digital automation (Integrately), retail (Atlas.kitchen), and employee engagement (PraisePal). Additionally, our e-commerce plugins have given us access to 2.75 million ecommerce stores (see e-store video (Handprint Tech, 2021)).

2.3. *Building a Moat*

Handprint is not alone in this space. Over the last few years, many ventures have popped up that are working in the impact integration space. A few notable names are Project Wren, Ecologi, Ecomatcher, Patch.io, all colibri, Nori, Cloverly, Earthly, Green Sparks, EcoCart, Carbon Click, and Klima. The key things that set Handprint apart from these other ventures are the following:

- Impact productisation with long-term monitoring and transparent financial accounting: you know where your money goes and how it is used.
- Impact diversity: not everyone wants to offset carbon emissions. A fast-growing portfolio ensures clients can find projects that align with their values.
- Impact as regeneration: carbon neutrality as an ambition level is insufficient to reverse global warming. Footprints (historical environmental guilt) rarely motivate sustained action. A focus on regeneration and planet positive action can inspire more people.
- Impact personalisation: Handprint establishes a P2P connection between clients and the impact projects they support.
- Impact co-ownership. Claims to the benefits created by a funding event are co-owned by the brand/platform and the consumer/user. This cements a long-term relationship between company and consumer: You are saving the planet together.

3. Advancing the Approach: Regenerate, Rewire, and Inspire

While most sustainability departments and ESG professionals focus on reduce, reuse, and recycle, the famous Lansink ladder of waste management, Handprint proposes a different approach summarised as 'regenerate, rewire, inspire.' The underlying thesis is that for most companies in the world, the singular focus on reducing negative externalities (which is now often boiled down to the singular goal of reducing carbon emissions) misses the point. The argument is that companies that are not making, moving, or mining physical goods have very little to do with how sustainability has been framed in the past decades. In the developed world, those companies account typically for less than 25% of employment (Schillebeeckx & Merrill, 2021). Even in the companies active in the making, moving, or mining of things, very few employees have a direct impact on the decisions that could improve sustainability credentials. Consequently, corporate sustainability is quite an exclusionary paradigm.

The focus on reducing negative externalities as the pinnacle of environmental responsibility has excluded companies in the service industry like consulting, accounting, legal, media, insurance, finance, as well as digital platforms and governmental and non-governmental organisations from taking meaningful sustainability action. The argument that these types of organisations typically have a small footprint and hence limited environmental responsibility starts from the misconception that the only environmental responsibility is the reduction of negative impact. Handprint argues that companies have two separate environmental responsibilities — one is to reduce their footprint and the other one is to grow their handprint. A company's handprint (positive impact) should not simply neutralise its footprint (negative impact). The story of footprint is one of limits and constraints. The story of handprint is one of abundance and growth. As Bill McDonough said in his famous TED talk on Cradle to Cradle, "The question is not whether growth is good or bad. The question is what do we want to grow?" And growing a bigger handprint is something we should all do.

3.1. *Regeneration*

Handprint's approach starts with a focus on regeneration. Regeneration goes "beyond sustainability and mitigating harm, to actively restoring and nurturing, creating conditions where ecosystems, economies, and people can flourish." (Stafford *et al.*, 2021).

Regeneration has three components: Reserve, Restore, and Rewild (Schillebeeckx & Merrill, 2021). **Reserve** argues that we need to maintain and expand our natural reserves. Currently about

11.7% of the planet's land and water bodies are protected as natural reserves. This needs to increase to 30% by 2030 as agreed in the 2023 Kunmig-Montreal Protocol. Doing so is not only possible and essential to biodiversity protection, it is also sensible from a purely economic vantage point (Waldron *et al.*, 2020). By expanding our natural reserves, we would simply be doing what smart households and fiscal conservatives have always known: You cannot keep eating away at your capital reserves indefinitely without repercussions. Building up our reserves needs to go hand in hand with recognising the rights of indigenous tribes living in them and providing space for nature to re-emerge.

Restore is a call for the active, human-led restoration of degenerated lands and waters. In the oceans, we need to restore fish populations, sharks, whales, and plankton to allow oceans to thrive again. We need to halt plastic disposal into the oceans and clean up beaches, rivers, and creeks to reverse oceanic plastic pollution. This will require massive investments in technology and a behavioural revolution. At once, we need to increase surveillance of fishing practices to ensure sustainable fishing is not simply a pipedream but a reality.

On land, we need a global forestry plan. There is great wealth to be created for the planet and for vanguard businesses that recognise this opportunity. For example, supporting mangrove reforestation in regions like Indonesia and Myanmar costs about 3,000 USD per hectare. The carbon sequestration alone (which typically is the reason companies would support this) is estimated to be about 1,000 tons per hectare, which at current prices for 'blue carbon' is worth between 8,000 and 60,000 USD. The value of ecosystem services provided by such a hectare could reach 190,000 USD (Romañach *et al.*, 2018). It is an enormous investment opportunity.

Rewild is the third piece of the puzzle. Popularised by George Monbiot's beautiful narrative of the Yosemite wolves (TED, 2013), rewilding requires the introduction of keystone species like wolves to reinstate a lost balance. But on top of that, it must also entail the withdrawal of humans. For us to truly rewild nature, we need to take our modern society out of the equation. Our inability to live in harmony in nature forces us to live in harmony beside nature. Singapore is an example of a country where large areas of its tiny landmass are simply not accessible to the public. Let the wilderness be the wilderness again as Sir David Attenborough pleaded.

Increasing our natural reserves, restoring nature, and rewilding nature are the three components of regeneration. Handprint carefully curates impact partners that are actively working on nature restoration and rewilding and works for them as a commercial agent. Through its digital tools, corporate and individual actors are empowered to make micro contributions to these projects.

3.2. *Rewire*

Convincing companies that making micro contributions to regeneration is a good idea is not self-evident. As much sustainability work focuses on reduction, the cost savings are often clear and predictable. Reducing energy consumption saves money. Improving product design to reduce material wastage saves money. Building more sustainably may still be more costly, but if you take into consideration the positive effect this has on talent attraction and employee health, it very quickly saves money as well. This is less obvious for supporting regeneration because regeneration involves direct costs and hence only makes business sense if it can be activated to increase value creation and capture. This is where rewire comes into play.

Rewire focuses on the blueprint of activities that create and capture value for companies. It starts from a deep investigation of the business wireframes. A business wireframe is a systematic analysis of a business's activities, with a clear focus on those activities that create touchpoints with key stakeholders such as employees, customers, and suppliers. You can think of wireframes as the blueprints of how a company actually implements its business model. While business models are commonly studied and very popular, business wireframes are a lot less commonly used but significantly more useful for entrepreneurs. However, developing your business wireframe is also a lot more work. In a business model, you would, for instance, state that your main sales approach will be cold e-mailing. However, the business model itself says nothing about how this is practically implemented. How does a company build a capacity to send cold e-mails? How does it identify prospects? How does it improve the opening rate of emails sent? How does it avoid getting blacklisted as a spam sender? Wireframes go to the very heart of practical implementation activities:

How does a company respond to an unsatisfied customer? How does it move money between different parties? How does a creative get published on Instagram or LinkedIn? What is the routine for responding to a client wanting to initiate a new process?

Wireframes describe in sufficient detail the processes and activities a firm engages in. They describe the nuts and bolts of a business in a way that is significantly more actionable than the business model, which provides only a high-level overview of how the company creates and captures value but tells us very little about how a company practically engages its customers or suppliers, or what the internal HR process is for a promotion. The higher actionability of wireframes comes at the cost of them being significantly more idiosyncratic.

Codifying every business wireframe is a lot of work and it is not recommended that companies go through this exercise. The focus should be on key wireframes that drive value or costs, have significant risk implications (like wireframes for compliance and cybersecurity), and that absorb a lot of time from key personnel. By writing out wireframes, a specific process like an HR promotion is broken down into various components. It helps people figure out which aspects of the work can be automated or outsourced, and which should remain within the purview of a specific person.

Handprint works with companies to digitally rewire key activities that affect customers and employees and imbue them with positive impact. Practically, this means the following:

(1) Identify which wireframes are key value drivers for the company's growth and success.

(2) Lay out the set of activities that constitute the wireframe (like a designer would design rough wireframes for a new UX) and create alignment with the client on the involved parties, tools, and key interactions.

(3) Choose which wires (interactions among tools and people) in the wireframe form the critical touchpoints between the company and the selected stakeholder.

(4) Embed regeneration in the wire.

The last step is where Handprint's digital toolbox and impact portfolio come into play. By curating a wide portfolio of diverse impact projects, companies can easily find a cause that aligns with their values. By digitising and productising impact projects into discrete units of measurable impact, clients can purchase impact in the same way they could purchase any other service good they need for their business. By building tools that enable automation and integration (e.g., APIs, SDKs, and dashboards), any organisational activity that leaves a digital fingerprint can be linked to a regenerative action commitment. The company pledges to make a (micro) contribution to a regenerative project (i.e., grow a handprint) every time a specific digitally observable action happens.

3.3. *Inspire*

The third step in the Handprint approach is Inspire and it is a critical one because it answers the question of why. While regeneration is absolutely needed to protect biodiversity, the climate, and thus human survival, the business imperative is less clear. Can companies truly benefit from imbuing some of their wires with regeneration? The answer is yes.

Evidence from across the world suggests that customers are increasingly voting with their wallets and that since the COVID-19 pandemic started in late 2019, more and more people have become environmentally aware and are trying to consume more consciously. The same is happening in B2B contexts where more and more companies, even those that previously were never challenged on their sustainability credentials, are now being asked to demonstrate their sustainability agenda. To give one example, marketing and media companies are evaluating the environmental impact of the ads they create and the carbon emissions associated with putting ads on phones. Embedding regeneration in key stakeholder touchpoints can be a powerful differentiator. For companies that already have philanthropic or CSR budgets, the opportunity is massive and the risk is very low.

Imagine a large e-commerce store with an annual philanthropic budget of 100,000 USD. In the past, its CSR team would spend the money on various causes and would try to communicate about this inside the company as well as on the website through its sustainability pages and in the annual report, but none of these channels seemed to be very successful in reaching a broad audience. Very few customers and very few employees were strongly engaged with the company's philanthropic activities because it just did not relate to their work.

Now, that e-commerce store decides to rewire various activities, specifically focusing on customer touchpoints and employee engagement. In collaboration with marketing and sales, it identifies that cart abandonment is a key challenge in e-commerce because many people window-shop and do not buy. It also finds out that sending follow-up emails to previous buyers is great to garner brief attention, but that the marketing team often lacks engaging content and hence resorts mainly to offering discounts. With HR, the company identifies promotions, work anniversaries, and evaluations as key activities that engage employees.

Using historical data, the e-commerce company knows it averages two hundred sales per day. If it embeds a positive impact worth 1 USD in each sale, it expects to spend 73,000 USD. By rewiring the online checkout process, the company could demonstrate a credible commitment that for every sale, 1 USD is rerouted to a positive impact project, say reforestation. This gives the customer an additional reason to buy the product. It also personalises the action for the customer who feels that his/her decision directly makes a positive impact. The customer has a sense of ownership. In A/B testing with an Australian sports brand, Handprint established that using its e-commerce plugin and demonstrating this credible commitment at the moment of checkout increased the checkout rate by 16%, a surprisingly strong business case. In addition, supporting reforestation in a specific region and having access to credible content from the project, and how the land is transformed from barren to luscious in the course of a few years, offer great opportunities for storytelling. Rather than offering discounts, marketing could send emails with stories about how the project is evolving, highlighting the approximate location of where the tree a particular client planted was and so on.

That leaves 27,000 USD the company could use for employee rewards for promotions, anniversaries, and evaluations. There are many ways in which this could be concretely implemented but the idea is simple. Reward employees with impact and let them choose from a curated list of impact projects what type of project they would like to see supported. Like the customer, the employee feels empowered and develops a sense of ownership of the impact which strengthens loyalty.

4. Concluding Thoughts

The trifecta of regenerate, rewire, and inspire is a powerful growth engine and a differentiator that companies regardless of industry and size can engage with to appeal to their key stakeholders in new ways. As the world is heating up, more and more people want to work for companies that are purpose-led and create a meaningful impact in the world. Especially for those organisations not active in the making, moving, or mining of things, regenerate, rewire, and inspire form the best business approach to create real shared value.

But Handprint is a living organism that, like every start-up, needs to evolve quickly to stay relevant and to stay ahead of the competition. In July 2022, Handprint published a 6-minute video with some of the most up-to-date content about what Handprint is and how it seeks to make a difference in the world (Handprint Tech, 2022). By the time you read this chapter, that video is likely to be outdated, but the associated YouTube channel will be a valuable source of information for those interested in learning more about this company.

References

Handprint Tech. (2021, March 2). How Handprint works. https://www.youtube.com/watch?v= APQ_5vbJLS8&t.

Handprint Tech. (2022, July 21). What is Handprint? | How to make an impactful business. https://youtu.be/XbzVdlucDfk?si=hgta6zct1phN CiPW

Merrill, R. K., Schillebeeckx, S. J. D. & Blakstadt, S. (2019). Sustainable digital finance in Asia: Creating environmental impact through bank transformation. https://www.dbs.com/iwov-resources/images/sustainability/reports/Sustainable%20Digital%20Finance%20in%20 Asia_FINAL_22.pdf.

Romañach, S. S., DeAngelis, D. L., Koh, H. L., Li, Y., Teh, S. Y., Barizan, R. S. R. & Zhai, L. (2018). Conservation and restoration of mangroves: Global status, perspectives, and prognosis. *Ocean & Coastal Management*, 154, pp. 72–82.

Schillebeeckx, S. J. D. & Merrill, R. K. (2018). Miracle mangroves: Funding of green shields in the Bay of Bengal. https://ink.library.smu.edu.sg/cases_coll_all/219/.

Schillebeeckx, S. J. D. & Merrill, R. K. (2021). Conflicting institutional logics as a safe space for collaboration: action research in a reforestation NGO. In G. George, M. R. Haas, J. Havovi, A. M. McGahan, P. Tracey (Eds.), *Handbook on the Business of Sustainability* (pp. 344–361). Cheltenham, UK: Edward Elgar Publishing Ltd.

Schillebeeckx, S. J. D. & Merrill, R. K. (2021). Regeneration first. https://handprint.tech/regenerationfirst.

Stafford, M., Yee, C.M. Cherian, E., Prendergast, M., Safian-Demers, E. & Tilley, S. (2021). Regeneration rising: Sustainability futures. https://www.wundermanthompson.com/insight/regeneration-rising. (pp. 7).

TED. (2023, September 9). For more wonder, rewild the world | George Monbiot. https://youtu.be/8rZzHkpyPkc?si=gzkSrE4x8hLU6tQ0

Waldron, A., *et al.* (2020). Protecting 30% of the planet for nature. https://www.campaignfornature.org/protecting-30-of-the-planet-for-nature-economic-analysis.

https://doi.org/10.1142/9789811293108_0009

Chapter 9

Enhancing the Sustainability DNA of Singapore's Gardens by the Bay Through Induction Training*

Kevin Cheong and Thomas Menkhoff

On 19 November 2021, John Tan, Principal Consultant of local sustainability management consulting firm Green Advocates, boarded a taxi to Gardens by the Bay (GB), feeling confident that GB's management team would endorse his suggestions to enhance the sustainability DNA of Singapore's renowned national garden through a refreshed induction training programme. As he fastened his seat belt, Tan reflected on the past few weeks during which he had analysed GB's green efforts, what GB was doing to further reduce its carbon footprint and co-create new sustainability solutions to combat climate change, and finally, incorporate them into GB's induction programme.

---

*This is a reprint of an SMU Teaching Case (SMU-21-0048) written by Kevin Cheong and Thomas Menkhoff, originally published by the Centre for Management Practice (Singapore Management University) in 2022. The editors gratefully acknowledge permission from SMU's Centre for Management Practice to reprint the teaching case in this edition.

Having spent over 20 years in human resource development and large-scale project manpower outsource management, Tan had vast experience in creating onboarding and soft-skills training programmes. He had become a fervent advocate of protecting the environment, often encouraging his family and friends to conserve energy by using fans rather than air-conditioning, reducing food wastage, and taking public transport, rather than driving. Tan had just recently left his senior human resource position in a multinational corporation to join Green Advocates. The GB assignment would be his first major project deliverable, and he was eager to exceed his client's expectations.

1. Sustainability at Gardens by the Bay

GB was a national garden spanning 101 hectares located at the southern tip of mainland Singapore, adjacent to the Marina Reservoir. Its genesis could be traced back to a 2006 international master plan design competition that sought world-class design ideas for it (more than 70 entries were submitted by 170 firms, from over 24 countries, including 35 from Singapore). Development works began in November 2007, and in October 2011, GB was opened to the public (Gardens by the Bay, n.d.).

The Gardens consisted of three waterfront gardens: Bay South Garden, Bay East Garden, and Bay Central Garden. The two conservatories, Flower Dome and Cloud Forest, replicated the cool-dry climate of the Mediterranean and semi-arid sub-tropical regions, and the cool-moist climate of the Tropical Montane region respectively. Together, they housed a diverse collection of plants from various parts of the world. Sustainably engineered, both conservatories applied several cutting-edge technologies for energy-efficient solutions in cooling, which enabled GB to reduce its energy consumption by approximately 20% in comparison to buildings that used conventional cooling technologies.

Another innovative icon of GB was the collection of Supertrees, which had been embedded with environmentally sustainable features. Photovoltaic cells could be found atop seven of these

Supertrees to harness solar energy, while others were integrated with the conservatories to serve as air exhaust receptacles. One particular Supertree at the Golden Garden had a chimney that released treated and clean vapour of the flue gas from the biomass furnace, in compliance with air pollution standards set by the National Environment Agency (NEA). The 18 Supertrees (ranging in height from 25 m and 50 m) further acted as vertical gardens and were covered with tropical flowering climbers, epiphytes, and ferns.

In line with the Singapore Government's vision of remodelling the country from a 'Garden City' to a 'City in a Garden', GB was aimed at enhancing the quality of life of Singaporeans by amplifying the presence of greenery in the city. In 2019, over 13.1 million resident and foreign guests visited GB.

Besides showcasing the best of the plant kingdom, GB was designed according to the principles of sustainability: generation of carbon-neutral electricity, ground cooling with chilled water systems and use of biomass waste, water management systems integrating aquatic ecosystems, and green Supertrees.

As GB's management embarked on the next phase of its sustainability journey and set its sustainability agenda, it pondered over the right approach to engage its staff on this journey. The management had agreed that every new and existing employee had an important role to play in the transition of GB to a garden of net zero carbon emission. An effective induction training of new and existing staff members was identified as an urgent challenge to maximise GB's contributions to sustainability and climate change mitigation. Key components included making sense of the Sustainable Development Goals (SDGs) formulated by the United Nations (UN), essentials of Singapore's new Green Plan 2030, and environmental, social and governance (ESG) concerns as sort of a guidance system for enhanced sustainable action.

2. Sustainability Challenges

After reviewing GB's current strategic sustainability efforts, ranging from energy conservation to the creation of a conducive environment for biodiversity to thrive (as shown in Exhibit 1), vis-à-vis the

Exhibit 1.   Sustainability in the Gardens

Underlying the concept of Gardens by the Bay were the principles of environmental sustainability. Much effort had been made to plan and design for sustainable cycles in energy and water throughout Bay South Garden.

(1) Energetics of the Conservatories
Comprising two glass biomes, the Conservatories replicated the cool-dry climate of the Mediterranean and semi-arid sub-tropical regions, and the cool-moist climate of the Tropical Montane region. They housed a diverse collection of plants that were not commonly seen in this part of the world, some of which were of high conservation value.

The conservatories were a statement in sustainable engineering, and they applied a suite of cutting-edge technologies for energy-efficient solutions in cooling. This suite of technologies

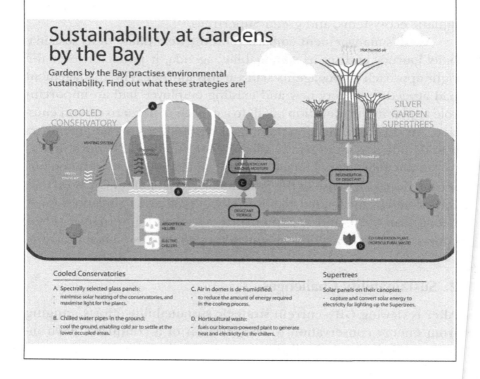

Sustainability at Gardens by the Bay

Gardens by the Bay practises environmental sustainability. Find out what these strategies are!

Cooled Conservatories

A. Spectrally selected glass panels:
- minimise solar heating of the conservatories, and
- maximise light for the plants.

B. Chilled water pipes in the ground:
- cool the ground, enabling cold air to settle at the lower occupied areas.

C. Air in domes is de-humidified:
- to reduce the amount of energy required in the cooling process.

D. Horticultural waste:
- fuels our biomass-powered plant to generate heat and electricity for the chillers.

Supertrees

Solar panels on their canopies:
- capture and convert solar energy to electricity for lighting up the Supertrees.

Exhibit 1.   (*Continued*)

allowed GB to reduce its energy consumption by approximately 20%, compared to buildings using conventional cooling technologies*.

*Third-party validated by an NEA-accredited Energy Services Company.

Minimising Solar Heat Gain
The two conservatories were fitted with glass with a special coating that allowed optimal light in for plants, but reduced a substantial amount of heat. The roof was fitted with sensor-operated retractable sails that provided shade to the plants when it got too hot.

Cooling Only the Occupied Zones
The Conservatories applied the strategy of cooling only the lower levels, thus reducing the volume of air to be cooled. This was achieved through displacement cooling — ground cooling by chilled water pipes cast within the floor slabs enabled cool air to settle at the lower occupied zone while the warm air rose and was vented out at high levels.

Generating Energy and Harnessing Waste Heat
Carbon-neutral electricity was generated on-site. At the same time, waste heat was captured in the process to regenerate the liquid desiccant. This energy co-generation was achieved with the use of a Combined Heat Power (CHP) steam turbine, which was fuelled by wood and horticultural waste from across Singapore. In doing so, the dependence on grid electricity generated from fossil fuels was reduced.

De-Humidifying the Air before Cooling
To reduce the amount of energy required in the cooling process, the air in Flower Dome was de-humidified by liquid desiccant (drying agent) before it was cooled. This desiccant was recycled using the waste heat from the burning of the biomass.

(2) Lake System: Dragonfly & Kingfisher Lakes
GB's lake system incorporated key ecological processes and functions as a living system. It acted as a natural filtration system for water from the

(*Continued*)

Exhibit 1.   (*Continued*)

Gardens catchment and provided aquatic habitats for diverse species such as fishes and dragonflies.

Encompassing two main lakes — Dragonfly Lake and Kingfisher Lake, the lake system was designed to be an extension of the Marina Reservoir. Water run-off from within GB was captured by the lake system and cleansed by aquatic plants before being discharged into the reservoir. Naturally treated water from the lake system was also used in the irrigation system for the Gardens.

The lake system depicted the role and importance of plants in the healthy functioning of GB's ecosystem. It raised awareness of the value that aquatic plants played in nature, and highlighted the significance of clean water in sustaining biodiversity.

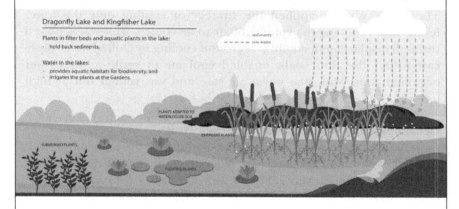

Filtering of Water Run-Off

Filter beds, comprising aquatic reeds and wetlands, were located where water entered and was discharged from the lake system. Water flow was reduced and sediments were filtered out.

Reducing Nutrient Load

Islands of aquatic plants and reed beds were incorporated to absorb nutrients such as nitrogen and phosphorus in the water. A reduction of nitrogen levels was critical to minimising alga bloom and ensuring better water quality.

Exhibit 1. (*Continued*)

Maintaining an Aquatic Ecosystem

Habitats for fish and dragonfly were created within the lake system by maintaining a diversity of aquatic plants, good water circulation and aeration. This kept in check potential problems such as mosquito breeding.

(3) Supertrees

(*Continued*)

Exhibit 1.   (*Continued*)

Environmentally Sustainable Functions of the Supertrees
Eleven of the Supertrees were embedded with environmentally-sustainable functions. Some had photovoltaic cells on their canopies to harvest solar energy for lighting up the Supertrees, while others were integrated with the Conservatories and served as air exhaust receptacles.

Energy Conservation Practice
As part of GB's sustainability effort to conserve energy, non-essential decorative and accent illuminations were switched off at night. Lighting at walkways and car parks was also reduced during hours when there were fewer visitors.

(4) Creating the Perfect Home for Biodiversity: Kingfisher Wetlands
Nestled amidst lush greenery and two of GB's prominent water bodies, Kingfisher Wetlands was a thriving ground for flora and fauna alike.

These freshwater wetlands were designed to enhance connectivity between the Lotus Pond and Kingfisher Lake, and at the same time segue seamlessly with the rest of GB. Concerted efforts were made to develop Kingfisher Wetlands sensitively, to allow ecosystem functions to recover and wildlife to reconnect quickly post-development.

Exhibit 1.   (*Continued*)

Planting a Carbon Sink

Over 200 mangrove trees and related plants could be found at Kingfisher Wetlands. These included native and critically endangered mangrove species such as the Firefly Mangrove (*Sonneratia caseolaris*) and Upriver Orange Mangrove (*Bruguiera sexangula*).

Mangroves were able to remove greenhouse gases from the environment and store up "blue" carbon to mitigate the effects of global warming. Interestingly, these plants were known to be able to sequester more carbon than rainforests. For this reason, mangroves were globally recognised as some of GB's best allies in the fight against climate change.

Submerged in water, mangrove trees also provided nursing environments and shelters critical to the survival of fish species, as well as animals like otters.

Providing Opportunities for Wildlife Encounters

The Wildlife Lookout at Kingfisher Wetlands created more opportunities for people to experience and encounter local fauna first-hand. Besides spotting some of the birds and animals that frequent the area from the Wildlife Lookout, visitors could also learn more about the biodiversity in GB from educational signage in the vicinity.

Creating a Conducive Environment for Biodiversity to Thrive

Among the distinctive elements of Kingfisher Wetlands were a series of water cascades and streamlets. Besides enhancing the area's aesthetics, these water features also promoted better aeration in the water, which in turn encouraged the growth of microhabitats where biodiversity could flourish.

Source: Gardens by the Bay.

UN 17 SDGs, as well as latest developments such as ESG integration, Singapore's 2021 SG Green Plan (refer to Exhibit 2) and the 2021 Conference of the Parties in Glasgow (COP26), Tan had met up with GB's management team to inquire about the practical people implications of its respective sustainability concerns. A Steering Committee led by the CEO and supported by the senior management team had been formed to devise measures needed to achieve

Exhibit 2.   SG Green Plan

Source: Government of Singapore, https://www.greenplan.gov.sg/resources/infographics.

The Singapore government recently announced the SG Green Plan for a "greener and more sustainable Singapore by 2030", which envisaged the achievement of several goals by 2030:

- City in Nature where every household will have access to a nature park within a 10-minute walk, and 1 million trees would be planted to absorb 78,000 tonnes of CO_2.
- Energy Reset where the city-state would transition to natural gas and solar power for its electricity generation, reduce energy consumption by 8 million megawatt-hours a year, and reduce domestic greenhouse gas emissions by 3 million tonnes a year.
- Green Economy through initiatives to develop enterprise capability in sustainability, transform Singapore into Asia's Green Finance centre, and drive sustainability research, innovation and enterprises.

Exhibit 2.   (*Continued*)

- Resilient Future with an aim to meet 30% of the city-state's nutritional needs through locally-produced foods.
- Sustainable Living with an aim to reduce waste landfills by 30%, tripling the population's active lifestyle, and including sustainability awareness at all levels of public education.

Source: Government of Singapore, https://www.greenplan.gov.sg/key-focus-areas/ overview.

the aspirations set for GB in the coming years. The team had a good idea of what it wanted to achieve. It was observed that with the challenges of daily operational demands, sustainability-related goals were not formally articulated and regularly communicated to the workforce at large. Also, the current induction programme for new hires was focused mainly on the sharing of GB's vision, mission and values, with periodic updates where relevant.

A substantial amount of information about GB's green agenda had been captured and published on GB's website, but Tan was unsure to what extent staff had imbibed them. Ongoing staff training was focused on job roles and task performance, such as delivering high quality guest services, managing wayfinding in the Gardens, learning how to operate new equipment or handling the upgraded ticketing system. Backend and operations staff were routinely sent for external courses to upgrade job-specific competencies such as data management, event management, and digital and social media marketing. Tan thus saw an opportunity to integrate the latest sustainability trends such as 'net zero' strategies or decarbonisation into staff training.

While walking around GB during earlier site visits, Tan had bumped into several operations staff and chatted with them. Many of GB's staff were botanists and horticulturists who cared deeply for greenery and biodiversity, and Tan recognised there were good prospects to help staff deepen their appreciation of how the larger sustainable goals were relevant to their work — from global goals to country goals to organisation goals and ultimately, to their work at the individual level. However, he felt that some staff might need more in-depth knowledge of the latest sustainability topics, and that not all staff were at the same level of sustainability awareness. Long-time employees might recall some details of current efforts and sustainability-specific aspects of the Supertrees, the cooled conservatories and GB's lakes, commenting that "it was a long time ago" since their induction. Some stressed that it would be beneficial to know the work of co-workers from departments better, and that perhaps more could be done to refresh and update their understanding of GB's "strategic" sustainability efforts. Overall, it appeared that staff were enthusiastic to learn more about sustainability and that there was a desire to be involved in GB's future green journey.

When asked how important sustainability was in their job roles, employees expressed their interest to learn more and to become green practitioners, because "Gardens by the Bay is a garden where sustainability is high up on its list of priorities". Some employees Tan had chatted with were already proactively reducing the single use of plastics and packaging by bringing their own reusable grocery bags and food containers for takeaways to work. Others felt that more needed to be done to make information and tips more "bite-sized" and easier to understand, rather than using technical jargon which was beyond the understanding of the general public.

Tan also discovered that operations and frontline staff were often asked by guests about GB's sustainability programmes, and how GB practised sustainability and environment protection. Some guests, especially school children, were keen to know more about the otters, insects and birds they had encountered during their visit, while some were keen to find out more about the air-conditioning systems of the two domes and how environmentally-friendly they were. Staff would do their best to provide answers and information, though there were occasions when they might find it challenging to engage with guests further as they did not have all the answers at the top of their heads.

Overall, Tan found that the staff's outlook on sustainability was largely aligned with GB's management, which had in recent times put in greater effort to drive sustainability at the Gardens through establishing its Sustainability Framework using the ESG model, as well as a Sustainability Steering Committee comprising management and staff across departments to steer and implement the Gardens' sustainability efforts in a holistic manner. Therefore, it was crucial to enhance the communication to staff so that all were on the same page in pursuing the common sustainability goals.

Tan also wondered how to keep employees enthused and engaged over a sustained period, especially employees who had been with the Gardens since the very beginning. "It cannot be a case of constant downloads from the top," he thought to himself, "there must also be uploads of ideas and contributions from the ground up too, with endorsement from GB's senior management."

As he met more employees, he also felt it would be beneficial to bring both new and existing employees together and level up their appreciation, understanding and knowledge of sustainable practices. There were some employees who were aware of sustainability concerns, some who were practising sustainability, and others who were already sustainability advocates and champions. "There is no 'one-size-fits-all' solution," Tan thought, "so what might the appropriate and differentiated approach be?" Would a differentiated induction training approach which segments trainees according to their awareness and skill levels make sense and be accepted by GB's top management? And how should such a programme be designed?

3. Issues and Questions

As Tan thought about the key recommendations he planned to propose to structure the induction training implementation plan, he pondered one more time over the key issues and questions that had come up in his review. How could GB communicate sustainability more strategically through its induction programme, when its emerging strategic Sustainability Framework was taken into consideration, vis-à-vis its current sustainability efforts which put emphasis on water and energy management? In what ways could GB further its sustainability efforts and features in line with best sustainability management practices pertaining to the reduction of carbon emissions, greater use of renewables, more sustainable food consumption within the Gardens, greener mobility, and others? How could GB better measure the extent to which it had attained its sustainability-related goals with regard to the relevant UN SDGs, Singapore's Green Plan, and ESG concerns in a manner that is comprehensible to staff? What measures could GB adopt to integrate more relevant sustainability/ESG goals and content into its induction curricula that would cater to both new hires and existing staff in an engaging blended learning format, combining both online instruction and classroom training?

Tan was aware that GB's human resources (HR) team would like to curate a new induction programme that would be engaging, motivating and innovative. Therefore, he had spent quite some time researching relevant online and technology-based tools and solutions that would be incorporated into his recommendations. He felt that his biggest challenge was to meet the learning needs of GB's diverse staff by offering cutting-edge sustainability content via a fun mode of delivery, and getting the buy-in from both GB's HR and management teams. As the taxi stopped in front of the Gardens' main entrance, Tan realised he had plenty to talk about and discuss during his hour-long presentation to GB's management team.

References

Gardens by the Bay. (n.d.). Our History. https://www.gardensbythebay. com.sg/en/about-us/our-gardens-story/our-history.html. (Accessed on February 2022).

Chapter 10

Successful Ageing in 'Smart' Singapore

Ma Kheng Min

1. Introduction

Among the developed nations in the world, Singapore has one of the most rapidly ageing populations. It is estimated that one in four Singaporeans will be aged 65 years and above by 2030 (Ang, 2021), and this figure will become one in two by 2050 (Siau, 2019). With the Total Fertility Rate (TFR) down to 1.1 per woman (Government of Singapore, 2021), Singapore's population is ageing, with the median age rising from 42.2 years in 2020 to 53.4 years in 2050 (United Nations, 2019). Singaporeans are also living longer, with life expectancy increasing to 81.3 years and 86.1 years for men and women, respectively (Ho & Huang, 2018). However, it was found that about a decade of their sunset years, specifically, 9.3 years for men and 10.9 years for women, might be spent in poor health (Ho & Huang, 2018).

Since the 1980s, the Singapore government had been developing plans and policies to help seniors, defined as people who are 60 years of age and above, 'to not just add years to life but add life to years' (Ang, 2021). The concept of successful ageing, defined by the five indicators of no major diseases, no disability, high cognitive function, physically fit and mobile, and active engagement with life (Subramaniam *et al.*, 2019), has been a focus area of the government for the past four decades.

Since the launch of its Smart Nation initiative in 2014, the Singapore government has injected billions of dollars and resources into harnessing technology to effect a transformation in health, transport, urban living, government services, and businesses.[1] 'Digital Society,' one of the three pillars of this Smart Nation initiative, focuses on equipping citizens to embrace technological solutions for a better quality of life. For seniors, this means successful ageing in a smart age-friendly city with all the possibilities of Information Communication Technologies (ICT), Internet of Things (IoT), Cloud of Things (CoT), and Advanced Artificial Intelligence (Skouby *et al.*, 2014).

1.1. *What is Successful Ageing?*

Normal ageing is a natural human development process that is characterised by a decline in physical, mental, and social functioning.

John Rowe and Robert Kahn defined successful ageing as having 'low probability of disease and disease-related disability, high cognitive and physical functional capacity, and active engagement with life' (Rowe & Kahn, 1997). There are five indicators that define successful ageing: no major diseases, no disability, high cognitive function, physically fit and mobile, and active engagement with life (Subramaniam *et al.*, 2019).

With life expectancies increasing by up to three months a year since 1840 (Kohn, 2016), Lynda Gratton and Andrew Scott (Gratton & Scott, 2016), authors of the book 'The 100-year life,' have argued that the traditional notion of the three-stage life comprising education, career, and retirement is no longer tenable. Instead, life would

---

[1] See https://www.smartnation.gov.sg.

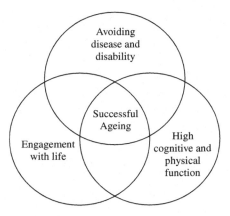

Figure 1. Model of Successful Ageing (Rowe & Kahn, 1997, p. 434)

consist of multiple stages peppered with a variety of work, entrepreneurship, continuing education, and breaks, and transitions.

In preparation for the 100-year life, the questions to ponder include the following[2]:

- How would you improve your long-term financial position?
- How would you support a longer working career?
- What would you do to improve your physical health and fitness?
- What would you do to upskill and upgrade your knowledge?
- How would you broaden and diversify your network?
- How would you seek a better work — life balance?

1.2. *What is Active Ageing?*

The World Health Organization (WHO) has defined active ageing as successful ageing coupled with positive experience (WHO, 2002). Specifically, 'active ageing is the process of optimising opportunities for health, participation and security in order to enhance quality of life as people age' (WHO, 2002, p. 12).

---

[2]See https://www.100yearlife.com.

Active ageing promotes the concept of productive living and living to the fullest potential according to one's needs, desires, and capacities. This suggests that people who are ill or have disabilities can experience active living while at the same time remaining active contributors to their family, friends, and communities.

Active ageing comprises the following four elements (World Health Organization, 2002, p. 13):

- *Autonomy* — the ability to make decisions and live according to one's rules and preferences.
- *Independence* — the ability to live independently with minimal help from others.
- *Quality of life* — 'an individual's perception of his or her position in life in the context of culture and value system where they live, and in relation to their goals, expectations, standards and concerns' (World Health Organization, 2002, p. 13). An individual's quality of life is determined by his or her ability to have autonomy and independence.
- *Healthy life expectancy* — living without disabilities.

According to the WHO, the three basic pillars to increase the rate of active ageing are health, participation, and security (WHO, 2002).

Singapore, like many countries in the world, has developed ageing policies based on the WHO's 3-pillar framework.

2. The Development of Ageing Policies

The issue of the ageing population has been on Singapore's national agenda since the 1980s. Several ministerial committees had been established to study the implications of an ageing population (see Exhibit 1).

In 1982, a 'Committee on the Problems of the Aged' was formed (Siau, 2019). The committee recommended several ways by which seniors could extend their contributions to the Singapore society

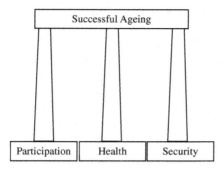

Figure 2. The Three Pillars to Drive Successful Ageing (WHO, 2002, p. 45)

Exhibit 1. List of 'Ageing' Committees and Their Recommendations and Policies

Year	Committee/Plan	Key Recommendations
1982	Committee on the Problems of the Aged	National policy to help seniors be physically and mentally fit for them to lead normal lives. Raising of the retirement age from 55 years to 60 years, and subsequently to 65 years.
1998	The Inter-Ministerial Committee (IMC) on the Ageing Population	Programmes to address the 'software'/services for seniors, the 'hardware'/provision of the necessary infrastructure and environment to support the growing number of seniors, and the 'heart wave'/ the attitudes, values, mindsets, and relationships of the Singaporean society. 'The Many Helping Hands' concept was launched to help seniors age well with the help of family and the community as the first and second lines of support, respectively, while the government would play its part by providing the necessary infrastructure and resources for seniors to age well.
2001	The Eldercare Master Plan	Strategy to develop senior-friendly community-based facilities as a part of the national infrastructure.
2004	The Committee on Ageing Issues (CAI)	Strategy to address the needs of the baby boomers (people born between 1946 and 1964), of whom many would be 65 years of age by 2010.

(Continued)

Exhibit 1. (*Continued*)

Year	Committee/Plan	Key Recommendations
2007	The Ministerial Committee on Ageing (MCA)	The Ageing in Place policy comprised four sub-policies focusing on providing senior-friendly housing; a barrier-free society that provided accessibility for seniors; a holistic, integrated, and affordable healthcare and eldercare system; and opportunities for seniors to pursue active lifestyles and well-being.
2011	The City for All Ages (CFAA) project	Building of senior-friendly communities.
2015	The Action Plan for Successful Ageing	Plans to provide opportunities (e.g., learning opportunities) and build a community and a city with senior-friendly infrastructure and transportation services. The plan comprised the three aspects that of proactive reach, preventive health, and active ageing programmes.
	Active Ageing Hubs	Development of Active Ageing Hubs to build communities that promoted volunteering, lifelong learning, exercise, and financial planning.
2017	Health clusters	Consolidation of health clusters to address the population needs in the west, central, and east of Singapore.

and economy. One recommendation of the committee led to the raising of the retirement age from 55 years to 60 years, and subsequently to 65 years (Goh, 2006).

In 1998, the Inter-Ministerial Committee (IMC) on the Ageing Population was set up to ascertain the 'software' or the programmes and services for seniors, the 'hardware' or the necessary infrastructure and environment to support the growing pool of seniors, and the 'heart wave' or the attitudes, values, mindsets, and relationships that underpinned Singaporean society (Goh, 1999). The collective approach of the 'Many Helping Hands' (Loo, 2017) was put in place to help seniors age well and successfully. The IMC also introduced the principle of 'ageing in place' that led to the development of elder-friendly homes, housing options, integrated community planning, and an increased budget allocation for the VWOs to provide healthcare and social services to seniors (Loo, 2017).

In 2001, the 'Eldercare Master Plan (FY2001 to FY2005)' called for a strategy to develop senior-friendly community-based facilities as a part of Singapore's 'national infrastructure' (Ministry of Community Development and Sports, 2001). Consequently, three Multi-Service Centres (MSCs) were established to provide one-stop community-based services to seniors and their families. The plan also saw the revision of the funding policy to help VWOs provide better services to seniors.

In 2004, the Committee on Ageing Issues (CAI) was formed specifically to address the needs of the baby boomers (people born between 1946 and 1964) — a generational group the oldest of which would reach 65 years of age in 2010 (Loo, 2017). Subsequently, the Ministerial Committee on Ageing (MCA) was set up in 2007 to coordinate the nation-wide efforts to translate the four policies that were developed by the CAI into reality (Loo, 2017). The four policies focused on developing and providing senior-friendly housing; a barrier-free society that provided accessibility for seniors; a holistic, integrated, and affordable healthcare and eldercare system; and opportunities for seniors to pursue active lifestyles and well-being.

In 2011, the 'City for All Ages' (CFAA) project was announced to build senior-friendly communities (Loo, 2017). Through a three-way collaboration, which included representatives from the community, public agencies, and community partners, the CFAA project raised awareness of the importance of health screening, promoted community programmes, and gathered input from seniors to ascertain their needs on health, community services, and urban development (Loo, 2017).

The 'Action Plan for Successful Ageing,' which was launched in 2015, led to an investment of S\$3 billion (US\$2.2 billion[3]) to help seniors age well and age successfully while leading healthy and active lives (Ministry of Health, 2016). The comprehensive plan comprised

---

[3]US\$1 = S\$1.35 on 29 July 2021.

Exhibit 2.   A Blueprint for Successful Ageing

60 initiatives covering 12 areas such as learning, employment, health and wellness, and housing

S$3 billion (US$2.2[4]) to kick-start the initiatives over 5 years

At the Individual Level	At the Community Level	At the City Level
• National Silver Academy to offer over 30,000 places for seniors by 2030 through 100 voluntary welfare and community organisations and post-secondary institutions. • National movement to promote senior volunteering — targeting 50,000 more seniors to become volunteers by 2030. • National Seniors' Health Programme to help at least 400,000 seniors age healthily through campaigns and workplace initiatives by 2030.	• Co-locating eldercare and childcare facilities at 10 new HDB Build-to-Order projects by 2025. • Passion Silver Card for every Singaporean aged 60 and above — offering priority queues and special discounts at participating stores.	• At least 10 Active Ageing Hubs in future HDB developments to provide seniors with daycare, learning activities and assist in their daily living, e.g., housekeeping and grocery shopping. • More senior-friendly transport infrastructure such as more intuitive signage and lifts at selected pedestrian overhead bridges by 2025. • Senior-friendly amenities in parks such as exercise equipment and therapeutic gardens. • S$200 million (US$149 million[5]) set aside to promote ageing-related research.

Source: Adapted from Ng (2015).

over 60 initiatives in 12 focused areas that included health and wellness, learning, and employment (see Exhibit 2). The plan focused on three thrusts that created the following:

- 'Opportunities for all ages': This was to help all residents including seniors continually learn and grow. The National Silver

---

[4]US$1 = S$1.34 on 21 October 2021.
[5]*Ibid.*

Academy[6] was set up specifically to provide a wide range of learning opportunities including 30,000 places (Huang, 2015) for seniors to stay active while learning and growing at the same time, and the National Seniors' Health Programme (Ministry of Health, 2015a) provided healthy lifestyle education and preventive health services.

- 'Kampong[7] for all ages': Kampong for all ages was an initiative that centred on inter-generational harmony. One outcome saw the provision of eldercare and childcare services within the same location in the new Housing and Development Board (HDB) developments to facilitate multi-generational interactions.
- 'City for all ages': This was Singapore's aspiration to be the model for successful ageing. A revised transportation plan that focused on senior-friendliness was introduced to address the needs of seniors in the MRT (mass rapid transit), buses, and on the roads.

Furthermore, a hefty sum of S$14.1 billion (US$10.4 billion[8]) was set aside to address the healthcare needs of the 'Pioneer Generation' (people born before 1949) and the 'Merdeka Generation' (people born between 1950 and 1959).

Prior to 2017, Singapore's public healthcare system was organised into six health clusters. In anticipation of increased healthcare needs, the six health clusters were regrouped into three in 2017 (Gene, 2021). The three clusters included the National University Health System (NUHS), National Health Group (NHG), and SingHealth to serve communities in the west, central, and east of Singapore, respectively. Within each health cluster, there was a medical school and a wide range of services offered including acute hospital care, primary care, and community care (see Exhibit 3).

---

[6]The National Silver Academy was a network of community-based organisations and post-secondary education institutions to offer learning opportunities to seniors. More information is available at: https://www.c3a.org.sg/faq.

[7]Kampong is the Malay word for village or society.

[8]US$1 = S$1.35 on 29 July 2021.

Exhibit 3. The Three Health Clusters

West Region of Singapore	Central Region of Singapore	East Region of Singapore
National University Healthcare System	**National Healthcare Group**	**National University Healthcare System**
• **General hospitals:** National University Hospital and Ng Teng Fong General Hospital.	• **General hospitals:** Tan Tock Seng Hospital, Khoo Teck Puat Hospital, and Woodlands Health.	• **General hospitals:** Singapore General Hospital, Changi General Hospital, and Sengkang General Hospital.
• **Community hospital:** Jurong Community Hospital.	• **Community hospitals:** Yishun Community Hospital and Woodlands Health.	• **Community hospitals:** Bright Vision Community Hospital, Outram Community Hospital, and Sengkang Community Hospital.
• **Primary care:** Bukit Batok Polyclinic, Choa Chu Kang Polyclinic, Clementi Polyclinic, Jurong Polyclinic, Queenstown Polyclinic, Bukit Panjang Polyclinic, and Pioneer Polyclinic.	• **Primary care:** Ang Mo Kio Polyclinic, Geylang Polyclinic, Hougang Polyclinic, Toa Payoh Polyclinic, Woodlands Polyclinic, Yishun Polyclinic, and Sembawang Primary Care Centre.	• **Primary care:** Bedok Polyclinic, Bukit Merah Polyclinic, Marine Parade Polyclinic, Outram Polyclinic, Paris Ris Polyclinic, Sengkang Polyclinic, Tampines Polyclinic, Eunos Polyclinic, and Punggol Polyclinic.
• **National specialty centres:** National University Cancer Institute, Singapore; National University Heart Centre, Singapore; and National University Centre of Oral Health, Singapore.	• **National specialty centres:** National Skin Centre.	• **National specialty centres:** National Cancer Centre, National Dental Centre Singapore, National Heart Centre Singapore, National Neuroscience Institute, and Singapore National Eye Centre.
• **Medical school:** Yong Loo Lin School of Medicine and National University of Singapore.	• **Specialised hospital:** Institute of Mental Health.	• **Specialised hospital:** KK Women's and Children's Hospital.
	• **Medical school:** Lee Kong Chian School of Medicine and National Technological University.	• **Medical school:** Duke-NUS Medical School.

Source: Adapted from Gene (2021).

3. Social Connectedness: Key to Successful Ageing

Ageing is a multi-faceted and complex issue that involves more than just seniors taking ownership of their health, e.g., adopting a healthy lifestyle. Relationships with families, friends, and communities should also be addressed as social connectedness is key to enabling older people to age 'successfully' and 'in place,' as well as forming the backbone of 'age-friendly societies' (Morgan *et al.*, 2021).

Social connectedness is 'the sense of belonging and subjective psychological bond that people feel in relation to individuals and groups of others' (Jetten *et al.*, 2017). Social connectedness is fundamental to active ageing because connectionist, along with group memberships, provides individuals with meaning, support, and agency, leading to a positive social identity. Thus, social connectedness provides a 'social cure' that is linked to health.

There are many benefits that social connections offer to seniors:

- *Longer life expectancy*: There are numerous studies that show that people with social connectedness, including time spent with friends, tend to live longer (Maier & Klumb, 2005). Those with a wider circle of friends also live longer (Glei *et al.*, 2005).
- *Better mental health*: Seniors with diverse networks, especially those with 'friends support quality' networks have fewer and reduced incidences of depression (Fiori *et al.*, 2006).
- *Social cure*: At the core of having a healthy life is social connections that exist and a sense of shared identity that these connections both produce and are produced by (Jetten *et al.*, 2017).
- *Preservation of cognitive function*: Seniors who participate in social activities experience less cognitive decline (James *et al.*, 2011), especially those who participate in social activities outside the family (Glei *et al.*, 2005).

4. Ageing in Place

The 2007 The Ministerial Committee on Ageing rolled out the Ageing in Place (Ministry of Health, 2007) policy to help seniors do the following:

- *Receive better healthcare at the right place*. The delivery of healthcare is being transformed from providing disease-specific and fragmented care to person-centric care. There is also a shift from hospital care to community care.
- *Live in senior-friendly homes and senior-friendly towns*. 'Continuing Care Precincts' including nursing homes and eldercare facilities have been included in the new HDB developments. The senior-friendly homes will have smart alert systems to monitor the safety and well-being of seniors. Towns will be made safer and more senior-friendly with the smoothening of pathways, installation of sheltered rest stops, etc.
- *Move about easily* with senior-friendly transportation network, e.g., wheel-chair-friendly buses and priority queuing system for seniors to get onto buses.
- *Enjoy public spaces such as parks*. Several parks have been redesigned, including relaxing gardens to provide horticulture therapy through the provision of contemplative and activity zones.

Additionally, three research centres for ageing have been established: e Geriatric Education and Research Institute (GERI), Centre for Ageing, Research, and Education (CARE), and National Innovation Challenge on Active and Confident Ageing.

Today, there are many more senior-friendly facilities, communities, and programmes than there were in the past for seniors to enjoy active and successful ageing. That said, there is more that can be done, such as supporting seniors who do not have the means to retire, especially those who are cash-poor but asset-rich.

5. Active Ageing Hubs

The 2015 Action Plan for Successful Ageing (Ministry of Health, 2015b) set out 70 initiatives in 12 areas to promote active ageing in an inclusive community coupled with the right infrastructure. A part of this infrastructure is the setting up of new communities called the Active Ageing Hubs. The Active Ageing Hubs are designed to promote volunteering, lifelong learning, exercise, and financial planning.

In 2018, the Agency for Integrated Care (AIC) was designated as the single agency to coordinate the delivery of aged care services, and to enhance service development and capability building across both the health and social domains. The Pioneer Generation Office was renamed the Silver Generation Office and joined the AIC that same year.[9]

The Silver Generation Office (SGO) is the outreach arm of the Agency for Integrated Care (AIC). Seniors volunteer as Silver Generation (SG) Ambassadors and provide caregivers and seniors with information about government policies and schemes on staying active and ageing well. They also connect them to services they need.

The AIC also supports partners in strengthening their capability to deliver quality care and brings partners together to meet the needs of the ageing population. There were 280 eldercare centres established to serve people in need, including seniors (Koh, 2020). These have now been reorganised as the Active Ageing Centre (AAC) or the Active Ageing Care Hub (AACH). By 2019, six Active Ageing Hubs have been opened (Begum, 2019). The AAC and AACH provide a suite of services which include active ageing activities (e.g., group exercise sessions, healthy cooking classes, health workshops, and social activities) to keep seniors engaged within the community; befriending services for vulnerable seniors who require additional social support (e.g., assisted living services such as housekeeping and grocery shopping); and referral services and informing seniors of government schemes, grants, and support. Centres also welcome seniors who would like to volunteer, such as helping with centre activities, visiting lonely seniors, or running errands for seniors. Active Ageing Care Hubs provide additional care services, such as daycare, dementia care, nursing services, rehabilitation, and management of emergency alert response calls.[10]

---

[9] See https://www.aic.sg.

[10] See https://www.aic.sg/care-services/active-ageing-centre-care-hub.

However, despite the extensive plans and efforts by the government to promote successful ageing, research revealed that only 25.4% of seniors in Singapore were experiencing successful ageing (Subramaniam *et al.*, 2019).

It was found that the odds of successful ageing of those aged 75–84 years is 0.3 times more compared to seniors aged 60–74 years old. This number dropped to 0.1 times for those aged 85 years and above (Subramaniam *et al.*, 2019). Ethnicity also affected the odds of successful ageing. The Malays and Indians experienced less successful ageing compared to the Chinese. And those with lower levels of education had lower rates of successful ageing compared to those who were more educated (Subramaniam *et al.*, 2019).

6. Challenges of the Ageing Community in Singapore

While longevity might be great, there are many challenges that an ageing community in Singapore faces:

- *Ageism*: There is a prevalence of ageism in the workplace. Ageism is more rampant in Singapore than among the other countries in Asia, with one in five workers reportedly experiencing age discrimination (Syed, 2020). It has been reported that older workers are often ruled out for even low-level jobs (Moore, 2017), and at least 50 ageism complaints are filed with Singapore's Ministry of Manpower every year (Iau, 2020).
- *Cash-poor but house-rich*: There is a variation of income and wealth among the third-agers. Those living the Housing Development Board (HDB) flats are more likely to be cash-poor but house-rich (Chan, 2016).
- *Longer employment to fund longer lives*: In Singapore today, one in four seniors still work (Liew, 2019). The employment rate of seniors has increased from 13.8% in 2006 to 26.8% in 2018 (Liew, 2019). Many seniors are choosing to work into their 80s and 90s to stay active. While some need money to survive, there are many who choose to work to be mentally and physically active (Ang, 2017).

- *War for talent*: With a sub-replacement fertility of less than 2.1 babies per woman (Gallagher, 2018), which is the critical number for population replacement, there will not be enough qualified young workers in years to come in Singapore.
- *Upskilling for the future of work*: Continuing education will become the norm as people reinvent themselves to have multiple careers throughout their lives. It is predicted that 47% of today's jobs will disappear in the next 25 years (Perry, 2016) as companies look to robotics, automation, artificial intelligence (AI), and algorithms to increase their competitiveness. The National Silver Academy of Singapore, which was launched in 2015, is aimed at helping seniors reskill and upskill. And more than 15,000 seniors have signed up for courses (*The Straits Times*, 2018).
- *Surge in healthcare costs*: The healthcare budget in Singapore has doubled from S$10 billion (US$7.3 billion[11]) in 2010 to S$21 billion (US$15.6 billion[12]) in 2018. And it is expected to triple to S$59 billion (US$43.7 billion[13]) in 2030.
- *Seniors experience declining health* including psychological, physical, and functional aspects of health compared to those who remain in the workforce (Grand & Chan, 2018).
  - Psychological health: loneliness, depressive symptoms, and cognitive function.
  - Physical: increase in chronic diseases including heart attack, cerebrovascular disease, and high blood pressure.
  - Functional health: decreased ability to perform physical activities.

Would ageing in a smart city alleviate some of these challenges?

7. Ageing in a Smart Nation

Singapore has been widely recognised as leading the way to becoming a world-class, tech-driven city-state. Singapore topped the 2021

---

[11] US$1 = S$1.35 on 6 October 2021.
[12] *Ibid.*
[13] *Ibid.*

IMD-SUTD Smart City Index for the third year running. Published by the Swiss business school Institute of Management Development (IMD) and the Singapore University of Technology and Design, the Smart City Index ranks 118 cities by how 'smart' they are. The Smart City Index takes into account input from the cities' residents of how technology has improved their lives. About 120 residents from each city were surveyed (Low, 2021).

There is the basic notion that smart cities with their information and communication technology (ICT) have great potential to create age-friendly living conditions, providing personalised healthcare, social services, and an age-friendly neighbourhood that can help overcome the mobility, visual, and cognitive challenges of older adults. ICT components like Internet of Things (IoT), Clouds of Things (CoT), and Advanced Artificial Intelligence (AAI) are key to developing age-friendly smart homes and smart cities.

Singapore has set its sights on becoming a world-class, tech-driven city-state. It is transforming itself into a Smart Nation, harnessing technology to transform how its people and businesses live, work, and play. The Smart Nation and Digital Government Office (SNDGO) is leading the transformation in key domains of health, transport, urban solutions, finance, and education in Singapore's journey as a Smart Nation. The SNDGO plans and prioritises key Smart Nation projects aimed at building long-term capabilities for the public sector and promotes adoption and participation from the public and industry to take a collective approach in building a Smart Nation.[14] Singapore has plans to build a Digital Society, Digital Economy, and Digital Government.

7.1. *Digital Society*

A Digital Society empowers every Singaporean with a fair chance of succeeding despite their differences or circumstances. It inspires them to dream bigger, as they get more interconnected with the world through technology.

---

[14]See https://www.smartnation.gov.sg/about-smart-nation/sndgg.

The aim is for every Singaporean to benefit from technology as part of a Digital Society. The Digital Readiness Blueprint[15] explains how the Singapore Government is making this happen:

- By making technology more accessible to every Singaporean.
- By improving Singaporeans' digital literacy.
- By helping the local community and businesses drive widespread adoption of technology.
- By designing inclusive digital services.

7.2. *Digital Economy*

A Digital Economy takes advantage of the latest technology to digitalise processes and drive business growth. This attracts foreign investments, which, in turn, creates new jobs and opportunities for those in Singapore. Singapore's pro-business environment, excellent tech infrastructure, close connectivity to major Asian economies, and the availability of investment place it in a good position to develop a strong Digital Economy.

7.3. *Digital Government*

A Digital Government is 'digital to the core and serves with heart.' Digitalisation is an effective means for the government to serve citizens with greater empathy by designing policies and services that are inclusive, seamless, and personalised for all.

In essence, a smart city provides a living environment where technology is seamlessly integrated in infrastructure and institutions. This should give seniors more opportunities to age successfully. Singapore's Smart Nation initiatives of Digital Society, Digital Economy, and Digital Government facilitate successful ageing as follows:

- **Access to high-speed internet for better connectivity**
There are plans for roll-outs of national NB-IoT networks and 5G mobile networks as well as the expansion of coverage of fibre

---

[15] See https://www.mci.gov.sg/en/portfolios/digital-readiness/digital-readiness-blueprint.

broadband and Wireless@SG to ensure that every business and person in Singapore has access to high-speed internet. This drives the widespread use of smart mobile devices, lowering prices of devices. Many seniors are using their digital devices and wearables like smart phones and smart watches on a daily basis to interact with smart services that monitor and manage their health and physical activities. Seniors who wish to age in place will also benefit from drone technologies, sensor technologies, voice command technologies, predictive analytics, and telemedicine.

Using digital devices can give seniors agency in their lives, reduce social isolation, and support social interaction with friends, families, and community. Bloomberg reports that seniors have become empowered using technology to overcome social isolation during the COVID-19 pandemic (Poon & Holder, 2020).

With an infrastructure that boasts of increased connectivity for all, seniors in Singapore will benefit from innovations in gerontechnology, such as social robots, smart homes, digital health monitoring, and wearables, that will address their desire to age in place, stay healthy, and be socially connected for successful ageing (Peine & Neven, 2021).

- **Better public transportation services and road use**

With the onset of mobility, visual, and cognitive impairment, concerns about safety can discourage seniors from leaving their homes.

Safe and accessible public transport is vital for seniors to move from place to place and continue connecting with people. Smart city applications like LED lighting controls and surveillance cameras enable the collection of data about traffic flow (human and vehicles) to create efficiency and improve sustainability in the use of public spaces.

The Yishun Integrated Transport Hub incorporates physical features that are senior- and dementia-friendly, including barrier-free boarding and alighting areas with priority queues and reserved seats, and prominent signage for easier way-finding.

Under the Silver Zone scheme, the Land Transport Authority (LTA) has implemented road safety features at selected residential areas to make it safer and more convenient for older pedestrians to cross the road. These include distinctive signs, traffic-calming measures, and road markings to reduce vehicle speeds and guide pedestrians to designated crossing points.

Technological advances in traffic and road management, traffic flow monitoring, live information sharing about road conditions, etc., enable seniors to reduce waiting time and better plan their day out.

- **More efficient public services**

During the COVID-19 pandemic, the Active Ageing Centre (AAC) or the Active Ageing Care Hub (AACH) had to deliver their services and activities remotely via digital platforms like WhatsApp, Telegram, Zoom, Microsoft Teams, and Webex Meetings. As a result, seniors and caregivers were able to enjoy the services from home and, in the process, learnt to use their digital devices to surf the internet for information and access other apps. This has enabled them to enjoy better and more efficient healthcare, social and community services, with significant improvements in day-to-day living.

- **Job opportunities**

A smart city will have many businesses and job opportunities. Businesses flourish in a smart city as they have access to an open data platform and real-time information for effective decision-making. A flourishing economy, in turn, means more jobs for Singaporeans, young and old. Seniors will have a chance at a second career in the 'smart' economy or just continue working in their jobs.

8. Closing Remarks

Singapore has been planning for an ageing population for 40 years. Ageing policies directed at Ageing in place and Active Ageing have cumulated in the proliferation of Active Ageing Centres and Active Ageing Hubs in all parts of Singapore to help seniors age successfully.

Ageing in Singapore is not without its challenges, but with the roll-out of the Smart Nation initiatives, the quality of life for the ageing population is on the mend and can only get better. We can look forward to a more inclusive workplace where 'smart' technology could facilitate the hiring of seniors seeking a second career or a new career in a different industry. Lifelong learning is set to rise with more seniors using digital devices to access learning content online or create content online. The widespread adoption of technology in 'smart' Singapore is a boost to society's aspiration to facilitate ageing in place. With an infrastructure that boasts of increased connectivity for all, seniors in Singapore will benefit from innovations in gerontechnology such as social robots, smart homes, and digital health monitoring devices that help them stay healthy and socially connected for successful ageing.

References

Ang, B. (2017, April 30). Age of golden workers: Many seniors working into 80s and 90s to stay active. *The Straits Times.* https://www.straitstimes.com/lifestyle/age-of-golden-workers.

Ang, J. (2021, March 23). Singapore's approach to healthy ageing is to see it as a positive force. *The Straits Times.* https://www.straitstimes.com/singapore/singapores-approach-to-healthy-ageing-is-to-see-it-as-a-positive-force?login=true.

Begum, S. (2019, April 24). First senior care centres, active ageing hubs operating 7 days a week in Ghim Moh and Telok Blangah. *The Straits Times.* https://www.straitstimes.com/singapore/two-senior-care-centres-and-active-ageing-hubs-operating-in-ghim-moh-and-telok-blangah.

Chan, J. (2016, December 23). Ageing isn't a tsunami. *Asia Research News.* https://www.asiaresearchnews.com/html/article.php/aid/10298/cid/2/research/people/singapore_management_university/ageing_isn%E2%80%99t_a_tsunami.html.

Fiori, K. L., Antonucci, T. C. & Cortina, K. S. (2006). Social network typologies and mental health among older adults. *The Journals of Gerontology Series B: Psychological Sciences and Social Sciences,* 61(1), P25–P32. https://doi.org/10.1093/geronb/61.1.P25.

Gallagher, J. (2018, November 9). 'Remarkable' decline in fertility rates. *BBC News.* https://www.bbc.com/news/health-46118103.

Gene, N. K. (2021, May 29). Three-cluster healthcare system should not be further reviewed: Ong Ye Kung. *The Straits Times.* https://www. straitstimes.com/singapore/health/three-cluster-healthcare-system-should-not-be-further-reviewed-ong-ye-kung.

Glei, D. A., Landau, D. A., Goldman, N., Chuang, Y.-L., Rodríguez, G. & Weinstein, M. (2005). Participating in social activities helps preserve cognitive function: An analysis of a longitudinal, population-based study of the elderly. *International Journal of Epidemiology,* 34(4), 864–871. https://doi.org/10.1093/ije/dyi049.

Goh, C. T. (1999, November 21). Speech by Prime Minister Goh Chok Tong at the launch of the senior citizens week on Sunday, 21 November 1999, at the Marina Promenade at 12:00pm. https://www.nas.gov.sg/archivesonline/data/pdfdoc/1999112101.htm.

Goh, O. (2006, October 1). Successful ageing — A review of Singapore's policy approaches. *Ethos, 1.* 14 October 2006. https://www.csc.gov.sg/articles/successful-ageing-a-review-of-singapore's-policy-approaches.

Government of Singapore. (2021). *Understanding Age-Specific Fertility Rate & Total Fertility Rate.* Department of Statistics Singapore. http://www.singstat.gov.sg/modules/infographics/total-fertility-rate.

Grand, H. L. Cheng & Chan, A. (2018). To work or not to work — Retirement and health among older singaporeans. *Centre of Ageing Research and Education (CARE), Duke-NUS Medical School,* 16.

Gratton, L. & Scott, A. (2016). *The 100-year Life: Living and Working in an Age of Longevity.* Bloomsbury Business.

Ho, E. L.-E. & Huang, S. (2018). *Care Where You are: Enabling Singaporeans to Age Well in the Community* (1st edn). Singapore: Straits Times Press Pte Ltd.

Huang, C. (2015, August 26). Government panel unveils key features of S$3b plan to help Singaporeans age well. *The Business Times.* https://www.businesstimes.com.sg/government-economy/government-panel-unveils-key-features-of-s3b-plan-to-help-singaporeans-age-well.

Iau, J. (2020, March 20). MOM penalises 5 employers for age discrimination in hiring practices. *The Straits Times.* https://www.straitstimes.com/singapore/manpower/mom-penalises-5-employers-for-age-discrimination-in-hiring-practices.

James, B. D., Wilson, R. S., Barnes, L. L., & Bennett, D. A. (2011). Late-life social activity and cognitive decline in old age. *Journal of the International Neuropsychological Society,* 17(6), 998–1005. https://doi.org/10.1017/S1355617711000531.

Jetten, J., Haslam, S. A., Cruwys, T., Greenaway, K. H., Haslam, C. & Steffens, N. K. (2017). Advancing the social identity approach to health and well-being: Progressing the social cure research agenda: Applying the social cure. *European Journal of Social Psychology*, 47(7), 789–802. https://doi.org/10.1002/ejsp.2333.

Koh, F. (2020, December 7). Seniors to receive help at 280 eldercare centres regardless of income and health: MOH. *The Straits Times*. https://www.straitstimes.com/singapore/seniors-to-receive-help-at-280-eldercare-centres-regardless-of-income-and-health-under-new.

Kohn, M. (2016, July 6). The 100-year life: How to make longevity a blessing, not a curse. *New Scientist*. https://www.newscientist.com/article/mg23130810-800-the-100year-life-how-should-we-fund-our-lengthening-lives/.

Liew, M. (2019, February 18). Seniors at work: The new norm for ageing in Singapore. *ASEAN Today*. https://www.aseantoday.com/2019/02/seniors-at-work-the-new-norm-for-ageing-in-singapore/.

Loo, D. (2017). Successful ageing in Singapore: Urban implications in a high-density city. https://doi.org/10.25818/16T3-84ZA.

Low, D. (2021, November 2). Singapore is world's smartest city for the third year: IMD Smart City Index. *The Straits Times*. https://www.straitstimes.com/tech/tech-news/singapore-is-worlds-smartest-city-for-the-third-year-imd-smart-city-index?login=true&close=true.

Maier, H. & Klumb, P. L. (2005). Social participation and survival at older ages: Is the effect driven by activity content or context? *European Journal of Ageing*, 2(1), 31–39. https://doi.org/10.1007/s10433-005-0018-5.

Ministry of Community Development and Sports. (2001). Eldercare master plan (FY2001 to FY2005). https://eservice.nlb.gov.sg/data2/BookSG/publish/1/166f6614-9ed2-4278-91ea-4be86caa64ec/web/html5/index.html?opf=tablet/BOOKSG.xml&launchlogo=tablet/BOOKSG_BrandingLogo_.png&pn=1.

Ministry of Health. (2007). How can I age-in-place. https://www.moh.gov.sg/ifeelyoungsg/how-can-i-age-in-place.

Ministry of Health. (2015a). Stay healthy. https://www.moh.gov.sg/ifeelyoungsg/how-can-i-age-actively/stay-healthy.

Ministry of Health. (2015b, August 26). $3 billion action plan to enable Singaporeans to age successfully. https://www.moh.gov.sg/news-highlights/details/3billion-action-plan-to-enable-singaporeans-to-age-successfully.

Ministry of Health, Singapore. (2016). *I Feel Young in My Singapore: Action Plan for Successful Ageing.*

Moore, H. (2017, February 28). It really is harder to find a job as you get older. *Ladders*, Business News & Career Advice. https://www.theladders.com:443/career-advice/new-study-proves-it-is-harder-to-find-a-job-as-you-get-older.

Morgan, T., Wiles, J., Park, H.-J., Moeke-Maxwell, T., Dewes, O., Black, S., Williams, L. & Gott, M. (2021). Social connectedness: What matters to older people? *Ageing and Society*, 41(5), 1126–1144. Cambridge Core. https://doi.org/10.1017/S0144686X1900165X.

Ng, K. (2015, August 26). $3b plan to help seniors stay active. *TODAY*. https://www.todayonline.com/singapore/ministerial-committee-ageing-announces-s3-billion-plan-help-seniors-age-actively.

Peine, A. & Neven, L. (2021). The co-constitution of ageing and technology — A model and agenda. *Ageing and Society*, 41(12), 2845–2866. https://doi.org/10.1017/S0144686X20000641.

Perry, P. (2016, December 24). 47% of jobs will vanish in the next 25 years, say Oxford University researchers. *Big Think*. https://bigthink.com/technology-innovation/47-of-jobs-in-the-next-25-years-will-disappear-according-to-oxford-university/.

Poon, L. & Holder, S. (2020, May 6). The 'new normal' for many older adults is on the internet. *Bloomberg.Com*. https://www.bloomberg.com/news/features/2020-05-06/in-lockdown-seniors-are-becoming-more-tech-savvy.

Rowe, J. W. & Kahn, R. L. (1997). Successful ageing. *The Gerontologist*, 37(4), 433–440.

Siau, M. E. (2019, August 13). Elderly to make up almost half of S'pore population by 2050: United Nations. *TODAY*. https://www.todayonline.com/singapore/elderly-make-almost-half-spore-population-2050-united-nations.

Skouby, K. E., Kivimäki, A., Haukiputo, L., Lynggaard, P. & Windekilde, I. M. (2014). Smart cities and the ageing population. *The 32nd Meeting of WWRF*, Marrakech, Morocco.

Subramaniam, M., Abdin, E., Vaingankar, J., Sambasivam, R., Seow, E., Picco, L., Chua, H., Mahendran, R., Ng, L. & Chong, S. (2019). Successful ageing in Singapore: Prevalence and correlates from a national survey of older adults. *Singapore Medical Journal*, 60(1), 22–30. https://doi.org/10.11622/smedj.2018050.

Syed, N. (2020, December 15). Ageism revealed as the most rampant form of bias at work. Human Resources Director. https://www.hcamag.com/asia/specialisation/diversity-inclusion/ageism-revealed-as-the-most-rampant-form-of-bias-at-work/241930.

The Straits Times. (2018, March 22). 15,000 seniors have benefited from National Silver Academy courses since 2015. *The Straits Times.* https://www.straitstimes.com/singapore/15000-seniors-have-benefited-from-national-silver-academy-courses-since-2015.

United Nations. (2019, July 17). Singapore: Average age of the population from 1950 to 2050 (median age in years) [Graph]. Statista. https://www-statista-com.libproxy.smu.edu.sg/statistics/378424/average-age-of-the-population-in-singapore/.

World Health Organization. (2002). *Active Ageing: A Policy Framework.* Second United Nations World Assembly on Ageing. https://extranet.who.int/agefriendlyworld/wp-content/uploads/2014/06/WHO-Active-Ageing-Framework.pdf.

Chapter 11

Towards More Inclusive Smart Cities: Reconciling the Divergent Logics of Data and Discourse at the Margins*

Jane Yeonjae Lee, Orlando Woods, and Lily Kong

1. Introduction: Why 'Inclusive' Smart Cities?

Over the past two decades, various smart city initiatives have developed and normalised a discourse that digital infrastructures can ameliorate existing urban problems. At the same time, critical scholarship has reminded us of the interlinkages between smart cities, neoliberalism, and dystopian techno-visions of the future (Karvonen, et al., 2018; Cugurullo, 2013, 2016; Kim, 2010; Shwayri, 2013; Shin, 2016; Hollands, 2008; Kitchin, 2015; Luque-Ayala & Marvin, 2015; Wiig, 2015; March, 2018). In response, there has been consideration

---

*This is a reprint of an article originally published in *Geography Compass* 1–12 (License No: 5486950945665). The editors gratefully acknowledge permission from the publisher (John Wiley and Sons) and the authors Jane Yeonjae Lee, Orlando Woods, and Lily Kong to reprint the article in this edition.

of how smart cities can be more inclusive so that they are no longer
planned and mobilised based on a profit-driven vision, but rather on
a contextualised one that is applicable to the actual lives of the citizens. Ideally, these visions are to be informed by the voice of the
public to be 'citizen focused' and 'community focused' (Cardullo &
Kitchin, 2018a; Datta, 2018).

Whilst there is an implicit value to smart cities being imagined
as inclusive, we need to ask why 'inclusive' smart cities are necessary
and desirable, and in which context and for what reason (March &
Ribera-Fumaz, 2016; Wiig, 2015). We contend that the phrase 'inclusive smart city' has often been misinterpreted as being inclusive of
every segment of the population, without considering people's
choice or having become aligned with the concepts of scale, economic growth and expansion. Instead, we argue that inclusive smart
cities should be evaluated as to whether they are equal and liveable.
Although inclusiveness in the sense of equity is not always central to
the policy rationale or functionality of smart cities, the questioning
is crucial for scrutinising the 'black box' of smart cities, the politics
of knowledge production (Fischer, 2000) and how pervasive and
manipulative data can be (Leszczynski, 2016) in the name of making
an 'inclusive' and liveable smart city. In this sense, there is a need to
high-light the knowledge politics embedded within the discourse by
questioning what participatory democracy and data transparency
really mean in the context of the smart city. By way of extension,
there is also a need to disentangle and acknowledge how the different levels of, and pathways to, power can shape, and re-shape our
existing interests and knowledges towards political decision making
(Fischer, 2000).

In this article, we contribute to the ongoing development of the
notion of 'inclusive smart cities'[1] in ways that better serve the purpose of recognising, critically understanding, and conceptualising
smart cities that are framed according to fundamental notions of

---

[1]Discussion around inclusiveness in urban planning is not something new.
However, it is our contention that in any given context, the definition of inclusiveness has been rather vague; hence further scrutiny is necessary.

equity, justice, and liveability. We refer specifically to the growing bodies of literature on data transparency (Kitchin, 2014; Burns *et al.*, 2018) which re-define the publicness of smart cities (Cowley, *et al.*, 2018) and humanise smart urbanism (Kitchin, 2019) through, for example, the lens of gender (Datta, 2018). These literatures address the unequal production of, and access to, big data and related infrastructures of smart cities. They also tackle the question of how smart cities perpetuate the reproduction of urban inequalities even within the imaginations of being 'smart' and 'inclusive.' By way of 'humanising' smart cities, these literatures must continue to push the boundaries of inclusiveness if they are to foreground the radical reconfiguration of the smart city as a public-led vision that can give a voice to otherwise silenced digital 'subjects.'

In what follows, we first outline a set of studies which examine smart citizenship. From there, we refine a number of themes that have emerged within the growing literature. Second, we move onto a set of studies that examine the 'actually existing' smart cities such as bottom-up and grassroots smart cities. In the final section, we discuss the limitations in the current discourse around inclusiveness and attempt to re-define what we mean by inclusive smart cities. We do so by introducing a more nuanced understanding of inclusion as a way of reconciling divergent realities of data and discourse at the margins. We conclude by discussing how the current literature alerts and navigates us to a pathway towards critical thinking about inclusive smart cities that are aligned with notions of equal rights and access, justice and liveability of citizens. We also suggest a new area for further research so that geographical scholarship can critique and contribute to creating inclusive smart cities in both theoretical and practical ways.

2. Smart Citizenship

'Citizenship' encompasses concepts of not only identification and belonging, but also power, control and politics (Cresswell, 2013). By adding the aspect of 'smart' to the politics embedded within citizenship, we can start to see the extent to which smart citizenship can be construed as an exclusive construct that can be divisive and potentially problematic. By taking this stance, we can problematise and question the exclusionary underpinnings that smart citizenship discourses make. We can also begin to understand why they are still far from being 'inclusive.' In this section, we discuss how 'inclusiveness' within the context of discourses of smart citizenship has diverged from the (implicitly more realistic) realm of public policy and practice. We outline a number of critical studies which scrutinise exactly how smart citizenship engagement has been put into action at various city and state levels and how it has shaped the policy and urban infrastructures involved (Cardullo & Kitchin, 2018a, 2018b; Datta, 2018; March & Ribera-Fumaz, 2016).

2.1. *Active Versus Inactive*

Smart citizenship generally refers to civic engagement via technology whereby cities can be co-designed and co-created in a civic, inclusive and transparent manner. It is also a policy vision wherein urban citizens are encouraged to play a key role in the collective production and administration of the city (Roy, 2001; Datta, 2018; Luque-Ayala and Marvin, 2015; Kitchin, 2019). Such visions of smart citizen engagement are laden with discourses of inclusion and transparency. These visions are aided by the 'publicness' of smart cities, enabled by the increasing availability of public data in supporting entrepreneurship as a means of revitalising the local economy (Evans *et al.*, 2015; Evans, 2016; Bulkeley *et al.*, 2016; Pollio, 2016; Crivello, 2015). For instance, Dowling *et al.* (2018) illustrate how a smart city in Newcastle, Australia, was rolled out through a collaborative effort by universities, young entrepreneurs and local agencies instead of by big companies.

Often, however, the reality is that big private companies are the key players, and when they roll out smart cities, questions surrounding data unevenness arise: whose perspectives are data collected from, who has access to the data, who determines what data to collect, and in whose interests. For instance, Cisco in Songdo and IBM in other smart cities have been variously criticised for bringing in technological infrastructures while taking away the local financial asset, without bringing much impact to the actual lives of the city's inhabitants (Hollands, 2008; March & Ribera-Fumaz, 2016).

The problem, therefore, is that while public data may be made available, that data, and the city it produces, is shaped by, and targeted at, the urban elite (Pollio, 2006; Shelton *et al.*, 2015; Crivello, 2015). The technologically illiterate and the poor risk being marginalised, and the digital spaces that emerge from smart citizenship act as a functional separation between 'sealed-off technological enclaves and leftover marginalised spaces' (Vanolo, 2014, p. 891). The 'digital divide' (van Dijk, 2006) is a central concern. That apprehension should become even more prevalent as cities are being shaped through technology and are producing digitally inclusive or exclusive spaces (van Dijk, 2006). However, the 'digital divide' is not just between those who have and do not have access to the smart technologies, and further thought is required and must be expanded in regard to the marginal and socially vulnerable. The division between active and inactive smart citizens in the current literature has been treated in a rather simplistic manner. There is a need to further interrogate autonomy and agency within the purview of participating and non-participating smart citizens, and one must determine under exactly what circumstances the division may exist.

2.2. *Competition and Neoliberal Citizens*

The narratives of 'participatory' and 'civic' engagement have increasingly become part of the marketing strategy undertaken by public officials who are advocates of technocratic advancement (Angelidou, 2014; Grossi & Daniela, 2017; Kourtit, *et al.*, 2017; Wiig, 2015; Young, *et al.*, 2019), but the reality of practice is also increasingly being questioned. For instance, a case study of smart citizenship as a new participatory mechanism in the city of Barcelona suggests that 'while the Smart City concept appeals to the central role of the citizen and boasts of opening up new participatory mechanisms, behind these strategies there is an increasing presence of big ICT, consultancy firms and utilities that search for new business opportunities' (March & Ribera-Fumaz, 2016, p. 826).

Despite attempts at making smart cities 'citizen-focused,' smart urbanism remains rooted in rational, functional and paternalistic discourses instead of social rights, political citizenship and the common good. In other words, within the inclusive smart city paradigm, the city not only brings services to its people but in so doing generates local business and economic prosperity in the long run. As a result, it creates 'neoliberal citizenship' (Cardullo & Kitchin, 2018b). Smart citizenship participation has also been critiqued as a 're-branding exercise' of smart cities (Kitchin, 2015) and even as an empty signifier (Cardullo & Kitchin, 2018a). Such criticisms are directed towards those technocratic public authorities and large technology companies who view the city as a market that can be managed and optimised via technology.

Smart citizenship engagement has been deployed as a strategy through which local government authorities can '(re)brand' their city as an inclusive smart city (Pollio, 2016). It is a common trend for local cities to compete for state funding to be the next 'citizen-focused smart city,' as exemplified by cities in India (Datta, 2018) and by Toronto (Bunce, 2004). This competitive tendency creates rivalry and sets sometimes unrealistic goals for creating an image that the city is citizen focused. Such a strategy is embedded within 'austerity' politics (Bramall, 2013; Schui, 2014; Blyth, 2015; Pollio, 2016) which further emphasise that funding and resources are always limited. Therefore, not everyone can benefit from the smart initiatives, but each and every city has to actively compete against others for state funding. Hence, to be desirable, the smart city initiative has to be economically sustainable. The 'civic engagement' and connection with the local community merely remain as 'nice to have' soft policies for an inclusive branding.

2.3. *The Myth of Smart Inclusiveness*

So far, we have shown that the idea that 'smart citizenship' can create 'publicness' and open up an array of opportunities for the local citizens remains a myth. The current literature explores how inclusive smart cities mainly exist in an ideological sense and how they have failed to materialise in the realm of actual public engagement. The logic of the (neoliberal) smart citizen is a 'global agenda driven from the North, and the hegemonic image of the technologically advanced city as the most desirable typifies the modernisation paradigm' (Ordendaal, 2018, p. 246). Datta (2018) argues that there is an even greater ideal vision that the quality of life in India's smart cities will be similar to that of the European standards. This standpoint highlights the aim of Indian smart cities as setting for themselves the boundaries of western modernity and urbanism. 'Smart citizenship thus emerges at the moment in the large-scale transformation of India's economy when a new individualist and consumerist identity has begun to take hold across different social groups' (Datta, 2018, p. 408). Datta (2018) goes on to argue that smart city urban development and planning in India, sparked by middle-class privilege, would inevitably produce an uneven distribution of networked spaces. The result would be a socially and physically uneven urbanism.

For now, 'smart citizenship' engagement is but a policy vision and a marketing strategy to create an image of the city being inclusive. That is to say, it is an online platform that passively waits for 'urban smart citizens' to share their local knowledge and vision for the future on behalf of their community. On the other end of the spectrum, real smart citizenship engagement can be exemplified through 'smart activism' or 'cyborg activism.' For example, one study shows how the marginalised people living in publicly owned land in Cape Town vocalised their 'Land for People not Profit' activism using social media and smart technologies (e.g., real-time photo essays and hashtags) to fight against the state's plan to sell their land (Ordendaal, 2018). The assemblage of advocacy, grassroots movements against urban gentrification and smart technologies results in a true 'participatory democracy' that is aligned with the notion of 'right to the city,' equality and justice. More recently, there has been a call to search for 'alternative modes' of digital development based on a study in Barcelona's grassroots smart cities whereby informal and organically developed networks of corporations, communities, and associations work towards forming a technological sovereignty movement which formed new arrangements of urban life and democratic decision making in the city (Lynch, 2019). Looking into such ongoing processes of alternative modes of digitalization goes beyond the well-established critique of smart cities.

In order for smart citizenship participation to be truly 'citizen-focused,' change needs to happen in a way that allows for the transfer of power to be meaningful for *all* citizens. Such power is rooted in the 'right to the city' debate and in ideas beyond the market-driven interests of smart cities (Cardullo & Kitchin, 2018b). Despite being implemented at the urban, local, and individual scale through citizenship engagement, studies indicate that the assimilation and diffusion of smart city initiatives, along with the actual technologies and decision-making processes, are still defined by institutions that operate at a larger scale. Kong and Woods (2018a) illustrate how smart urbanism is 'plagued' by the paradox between access and choice. That is because 'smart urbanism only works if its key beneficiaries — the inhabitants of a city — choose to engage with the enablers of a smart lifestyle' (691). At this point, the discourses around smart citizens seem to undermine the inclusive smart city initiative. That is because they are dominated by exclusionary and divisive concepts that only heighten the already existing social inequalities. Below, we discuss how these inequalities may be reconciled through bottom-up smart cities in a way that can create new opportunities for inclusion.

3. Grassroots and Bottom-up Smart Cities

In expanding the discourse around inclusiveness, the current scholarship has turned attention to questions like: What about the really marginalised segment of the population that is not even part of this growing discussion of the public/private, global/local, digital/human praxis? What about the people who cannot afford to purchase a smart home gadget or a smart phone or a computer and who do not have the 'digital right' (Datta, 2018) to the city — not by choice but by their circumstances? What about the disabled person who would never be physically able to enjoy equal access to the smart-cycling infrastructure that may be owned by the 'public'? Social polarisation within the digital future is often positioned within the 'looming background' of economic growth and efficiency of smart cities (Hollands, 2008; March, 2018; Sterling, 2018; Tonkiss, 2018). In this section, we turn to some of the emerging literatures on inclusive smart cities which are focused on either helping or serving the social needs of the marginal segments of the population.

3.1. *Actualising 'Inclusive' Smart Cities at the Margins*

The logic of redefining, materialising, and 'actualising' (Shelton *et al.*, 2015) is to step away from being too critical and thus hollowing out the actual notion of the smart city. The idea is to understand the complexities behind smart cities as assemblages of different political agendas, ideologies and actors that are in place to push for positive technological impacts on the lives of citizens and for the good of society (Kitchin, 2019; Shelton *et al.*, 2015). It is about understanding the newly emerging 'smart practices' in a tangible and intangible manner (Neirotti *et al.*, 2014) and actively searching for what 'smart cities produce or allow to be done differently [and] how [that] reconfigures the urban' (Veltz *et al.*, 2018, p. 134). While cities such as Songdo and Masdar have failed to implement their original masterplans fully, scholars have argued that it is important to understand how the smart city paradigm, as a policy and new ideology, has become 'awkwardly integrated into the existing social and spatial constellations of urban governance and the built environment' (Shelton *et al.*, 2015, p. 14).

Several studies have explored such integration. For example, in China, Lin *et al.* (2015) argue that while the mode of e-governance has produced a social sustainability, thereby forming the basic relationship among the state, market and society, rural migrants and other marginal social groups are excluded. Using migrant communities in rural areas in China as a case study, they introduce a new web-based planning support system that can generate smart governance for people at the margins (Lin *et al.*, 2015). Another study of migrant communities in Toronto (Rosu *et al.*, 2017) introduces a new web-based service which is designed to give support to low-income migrants. For instance, the web connects middle-income families who may want to discard their home goods with the low-income migrants who may be in need of such items but cannot afford to buy them. Through a complex computational system and using artificial intelligence based on the large dataset, the web service promises to connect the people who are willing to share their goods in a more efficient manner than the currently existing web services can provide (Rosu *et al.*, 2017). Those types of studies tend to be university-led projects, often initiated by schools with a technological background and extensive funding sources (Howe & Fox, 2017; Hussain *et al.*, 2015; Kim *et al.*, 2015). For example, the aging society is a pertinent issue globally (Moser, 2013; Trencher & Karvonen, 2018), and in Singapore, a project has been initiated by Singapore Management University to develop smart home technologies that can help assist elderly individuals who are living alone at home (Goonawardene *et al.*, 2018; Kong & Woods, 2018b).

However, university-led innovations tend to remain as pilot research projects and never evolve into more. They also may not be solving the 'real problems' because they are driven by technological interests without considering the social and political factors or the actual end-user. It is uncertain how many people would actually use these new services. It is also unclear how big the potential is for them to make meaningful partnerships and effective collaborations with various private and public stakeholders to carry out their socially inclusive smart services at the city scale. It may take many actors from higher education institutions, venture capitalist firms, technology companies, and governments as well as more tacit knowledge for university-led innovations to bring a meaningful impact to bear and to become a catalyst for change (Addie *et al.*, 2018). To this end, Woods (2020, pp. 5–6) recently argued that, in order to 'foster inclusion from the outset,' there is a need to develop 'solutions that do little things that help to augment, and incrementally improve, pre-existing patterns, processes and paradigms' of urban life.

The point here is that attempts to foster inclusion within the smart city should focus on small and incremental changes for the better, rather than transforming the urban in new, unpredictable, and potentially divisive, ways.

3.2. *Prescription and Temporality*

In the processes of 'actualising' smart cities at the margins and creating 'do it yourself' smart city innovations, conflicts of prescriptive logics, and rhetoric of temporality arise. A lot of the social innovation rhetoric can be found in the grassroots-led smart city initiatives. For instance, a study by Shelton *et al.* (2015) shows that the local government authorities in Philadelphia have been undertaking a digital-inclusion effort to provide Internet and technical education to the low-income and digitally illiterate residents. In doing so, they armed these individuals 'with the skills to be competitive for jobs in the 21st century information economy' (Shelton *et al.*, 2015, p. 19). What the study found, however, is that the jobs targeted for offering to these low-digital-literacy residents were located far from the poor neighbourhood. Hence, even if the marginalised people were trained and offered these 'technical' jobs, their everyday commute was too difficult to enable them to actually accept the job offer. This tendency shows that the project aimed at lessening the 'digital divide' missed the 'bigger picture.' Such an oversight cannot be solved through technology. Moreover, sometimes a lot of the bottom-up movements seem to just 'scratch the surface' without making much impact on governmental intervention (Shelton *et al.*, 2015). That is because authoritative instrumentalism and a paternalistic view still exist which strive to produce a 'known,' desired outcome (Shelton *et al.*, 2015; Shore & Wright, 2011).

Critical scholarship within geography alerts us that 'making smart cities for the poor' is not always the right answer. That is because such cities will only solve short-term problems. Hence, in the name of making 'inclusive smart cities,' the project may 'thrive for a few years and then fade away due to lack of funding or interest by residents' (Trencher & Karvonen, 2018, p. 272). At their current stage, the small innovation projects aimed at reducing social inequality and directly improving the lives of the vulnerable population are seen as an 'alternative' option. Such an option is not aligned with the state government's smart rationalities of making an economically viable, self-sufficient city. These community engagement projects remain as 'nice to have' soft policies and may dissolve once they are branded as an 'inclusive smart city.' Then the majority of funding may get used elsewhere instead of on the people who are actually marginalised.

Furthermore, geographers argue that the projects and initiatives targeted specifically towards the socially and economically marginalised must be free of civic paternalism and stewardship, wherein companies and city authorities decide what is best for their citizens and deliver on behalf of citizens (Cardullo & Kitchin, 2018b). Instead, 'a larger understanding of urban infrastructure systems is necessary to move from data to information to knowledge and, ultimately, to action for urban sustainability and human well-being' (Ramaswami *et al.*, 2016, p. 941). In other words, the existing urban inequalities of uneven mobility, uneven housing and living conditions, and unequal access to certain infrastructure and services based on people's class, gender, ethnicity, race, and income must be addressed first. This must happen before the 'smart intervention' takes place.

4. Towards More Inclusive Smart Cities

Overall, there appear to be divergent logics of speed and scale when it comes to citizen participation versus smart city initiatives. The former is slow, bottom up and 'inclusive,' and the latter is fast, top-down and 'exclusive.' Yet, as we have discussed so far, such a dichotomy is arbitrary. That is because many of the currently existing 'inclusive,' bottom-up smart cities are also conflated with the rhetoric of being top-down, generating neoliberal spaces, and are complicated by knowledge politics (Fischer, 2000). Instead, the question should be on how can we reconcile them in a way which many of the existing studies fail to capture. In this final section, we provide further critique of inclusive smart cities and propose ways to move forwards in achieving a more nuanced and critical understanding of inclusive smart cities. We propose that this be achieved, first, by redefining the discourses and epistemologies of 'inclusiveness' in smart cities. Second, it can be accomplished by engaging more critically with the studies of representations and knowledge politics of data in an attempt to 'know what we are criticizing.' In this way, one can reconcile the divergent realities of data and discourse at the margins.

4.1 Redefining the Discourse: Emphasising Autonomy on the Margins

There are a number of limitations in the current literature that are concerned with social inclusion of the marginalised and socially vulnerable population in the smart cities narratives. First, the assumption is that an economic logic underpins the 'digital right' to the city. Such a logic may be the primary factor, but it's not the only one. As mentioned at the beginning of this article, the division between the active and non-active smart citizens has been treated in a rather simplistic manner. As such, there is a need to further interrogate autonomy and agency within the purview of participating and non-participating smart citizens. There is also a need to investigate under exactly what circumstances this division may exist.

Citizens may choose to be digitally marginal as an act of resistance or subversion. In terms of eldercare, for example, choosing not to engage with smart technologies is often based on elderly peoples' preference to be 'cared' for in more proximate, more relational and more human-centred ways instead of being cared for by smart technologies (Kong and Woods, 2018b). With such insight, research could flesh out some of the assumptions and biases that are embedded within normative understandings about the 'margins' and the 'vulnerable.' The current discourses around inclusive smart cities turn to the marginal groups as a homogenous group that is vulnerable and stress that all marginal people should be included. Future studies can focus on interrogating what is meant by a marginal group and defining who is socially, psychologically, economically, physically, and emotionally marginalised or left out from the smart city's growth — and also how they can be spatially and temporally marginalised. Such plurality of 'margins' can also push us to clearly define why we should be considering such individuals. As discussed here, these refinements have not been achieved in the current discourse of inclusive smart cities. Furthermore, studies fail to define who we should be inclusive of and why. In rectifying this situation, we can begin to re-think the notion that inclusion in a 'smart' city is inherently good or desired. As critical scholarship has suggested, this assumption can be problematic. One needs to identify and explain the (potential) benefits of a fully realised smart city and clarify why inclusion is necessarily desirable.

4.2. *Reconciling 'Data' at the Margins*

Collecting data about the city in order to make better urban and policy decisions has existed for many years. However, within the context of smart cities, 'data' means large amounts of data collected through ICT technologies and sensors that are then analysed by machine learning and AI that signifies the 'objectivity' of data. Hence 'data' in the context of smart city has been associated with urban efficiency, better insights, and better planning and objective policy making. It is within the critical understanding of data analytics that one can start to critically think about the role/s of data in 'fully' realising a smart city — from collection, to analysis, to response. Such a perspective can also allow researchers to gather the potential for inclusion/exclusion to be reproduced and/or overcome in the data assemblage of smart city making. Critical geographers have been at the forefront of challenging the 'objectivity' of big data and have commented on its manipulation of and situatedness in smart city knowledge politics. They have also stressed the need to bring 'data transparency.' By way of looking at the data through a critical lens, we can problematise the fact that while the city becomes more transparent or predictable, the problem lies in the fact that it is also liable to be manipulated by those who collect and use the data (Ash *et al.*, 2019; Kitchin, 2014).

A closer examination of data alerts us that smart cities are reproducing urban inequalities in a new way. This is not a novel claim, but as more and more smart cities are being rolled out with the next new ideal in urban planning initiatives, technological interventions are rationalised as the persistent norm in producing uneven geographies. For instance, in the process of sharing big data via public-owned entities and using certain smart technologies to map out city information, some neighbourhoods either get represented in a subjective manner (e.g., 'unsafe') or are ignored and left off the map completely. Studies argue that the reliance on big data to predict and control the future city diverts policy makers to the idea that cities must be securitised to prevent them from 'urban risk.' Such 'big data security assemblage,' however, is often uneven, producing greater urban inequalities (Datta, 2018; Kitchin, 2014; Ordendaal, 2018; Shelton *et al.*, 2015).

By way of example, a 'safety app' named 'Ghetto Tracker' is framed by racist thinking. It produces an explicit image which designates a certain neighbourhood as 'unsafe.' Such a label is 'highly speculative in orientation' (Leszczynski, 2016, p. 1698). The discussions around persuasive data are echoed by other scholars of digital geographies. These researchers attempt to develop a critical understanding of big data by emphasising (1) discussions around the politics of knowledge production, (2) *representations* of data technologies, and (3) the embeddedness of political and economic power relations within big data knowledge production (Burns *et al.*, 2018; Elwood & Leszczynski, 2018; Fischer, 2000; Halphern, 2014; Kitchin & Kneale, 2001). Recently, a study of grassroots smart cities in Spain has shown how citizens at the margins are empowered by various collective digital grassroots innovations (Boni *et al.*, 2019). Based on four different cities in Spain, the study illustrates that Spanish citizens now can openly initiate their own networks and infrastructure via an open network system, which is supplemented by an open data system that offers transparency in the collection, analysis, and usage of data. Those new digital grassroots movements allow collective actions in searching and solving new shared problems. In this way, one could imagine 'innovation' not just in a materialistic form, but also symbolically through the new construction of citizenship and democracy (Boni *et al.*, 2019).

4.3. *Directions for Future Research*

While there has been a growing discussion on critically understanding the rationalities and functionalities of smart cities and on deconstructing the data assemblage, the methods of *how* to study smart cities have received relatively less attention. Marvin *et al.* (2015) have suggested, 'the ways in which the social, economic and political potential of smart urbanism is fundamentally produced with and through technologies remain beyond the reach of social science perspectives' (p. 3). Indeed, the social science perspectives often rely on methods such as tracing media, document and policy analysis and interviewing a wide variety of involved stakeholders. They do so as a way of 'tracing' and 'following' the processes and politics involved in the city's smart technology rollouts and data assemblage. These methods have their own merits and allow deep understandings to be gained about the causality and processes of smart cities or the end-user's experiences of living in and/or with the smart cities technologies. However, this means that they are ways of 'gazing' upon smart cities from the outside (Hollands, 2008; Marvin *et al.*, 2015; Ordendaal, 2018). As Marvin *et al.* mention, they do not allow social scientists to understand or untangle the 'fundamentality' of technologies — or in the context of smart cities — to obtain deep understanding of the data analysis itself. Geographers have already begun this conversation and have seen an important relationship and overlap between software studies and GIS (Lally & Burns, 2017). They have called for critical geographers to work with the data and software GIS studies as a means of moving beyond the criticism and working together to produce critically engaged software technologies (O'Sullivan, 2019).

We argue that limited tools and skills are available within the social sciences to fully grasp what is happening within the 'black box' of smart cities. There is limited methodology in the social science arena. Other than 'gazing' upon smart cities from the outside (Hollands, 2008) through document analysis and interviewing different stakeholders who are involved in smart cities, the current work needs to become truly interdisciplinary. Only then can it fully tackle the epistemological and technical understanding of the data-learned and data-knowledgeable cities. Perhaps this would require an ethnography of the data scientists who are directly involved in making smart cities to learn their struggles, so as to bring humanistic expertise to the practical world of data collection (Neff *et al.*, 2017). Or, one could imagine a group of psychologists, behavioural scientists, geographers, historians, and data scientists working together to answer how data could be used to solve certain social problems. Perhaps such interdisciplinary work might be too difficult for projects that are short-term or with limited resources, yet it is having certain kinds of 'algorithmic sensibility' on the humanistic side, and gaining humanistic and contextual mind-sets on the data scientist side that could be a beginning of a truly interdisciplinary project. In this way, critical scholarship can begin to not only critique but also contribute to reducing the continuous urban inequalities in which smart cities and the current indulgence over big data seem to be reproducing.

Acknowledgments

This research was funded by 'Singapore Management University Societal Challenge Capacity Building Fund (Grant Number: SC-RCB-2018-003)'. The authors would also like to thank the anonymous reviewers.

References

Addie, J., Angrisani, M. & De Falco, S. (2018). University-led innovation in and for peripheral urban areas: New approaches in Naples, Italy and Newark, NJ. *European Planning Studies*, 26(6), 1181–1201.

Angelidou, M. (2014). Smart city policies: A spatial approach. Cities, 41, S3–S11.

Ash, J., Kitchin, R. & Leszczynski, A. (2019). *Digital Geographies*. Los Angeles, CA: Sage Publications, pp. 270–280.

Blyth, M. (2015). Austerity: *The History of a Dangerous Idea*. Oxford, UK: Oxford University Press.

Boni, A., López-Fogués, A., Fernández-Baldor, Á., Millan, G. & Belda-Miquel, S. (2019). Initiatives towards a participatory smart city. The role of digital grassroots innovations. *Journal of Global Ethics*, 15(2), 168–182.

Bramall, R. (2013). *The Cultural Politics of Austerity: Past and Present in Austere Times*. Basingstoke, England: Palgrave Macmillan.

Bulkeley, H., Coenen, L., Frantzeskaki, N., Hartmann, C., Kronsell, A., Mai, L., Marvin, S., McCormick, K., van Steenberg, F. & Palgan, Y. V. (2016). Urban living labs: Governing urban sustainability transitions. *Current Opinion in Environmental Sustainability*, 22(C), 13–17.

Bunce, S. (2004). The emergence of 'smart growth' intensification in Toronto: Environment and economy in the new official plan. *Local Environment*, 9(2), 177–191.

Burns, R., Dalton, C. M. & Thatcher, J. E. (2018). Critical data, critical technology in theory and practice. *The Professional Geographer*. 70(1), 126–128.

Cardullo, P. & Kitchin, R. (2018a). Being a 'citizen' in the smart city: Up and down the scaffold of smart citizen participation in Dublin Ireland. *GeoJournal*, 84, 1–13.

Cardullo, P. & Kitchin, R. (2018b). Smart urbanism and smart citizenship: The neoliberal logic of 'citizen-focused' smart cities in Europe. *Environment and Planning C: Politics and Space*. 37(5), 813–830.

Cowley, R., Joss, S. & Dayot, Y. (2018). The smart city and its publics: Insights from across six UK cities. *Urban Research & Practice*, 11(1), 53–77.

Cresswell, T. (2013). Citizenship in worlds of mobility. *Critical Mobilities*, 2013, 105–124.

Crivello, S. (2015). Urban policy mobilities: The case of Turin as a smart city. *European Planning Studies*, 23(5), 909–921.

Cugurullo, F. (2013). How to build a sandcastle: An analysis of the genesis and development of Masdar city. *Journal of Urban Technology*, 20(1), 23–37.

Cugurullo, F. (2016). Urban eco-modernisation and the policy context of new eco-city projects: Where Masdar city fails and why. *Urban Studies*, 53(11), 2417–2433.

Datta, A. (2018). The digital turn in postcolonial urbanism: Smart citizenship in the making of India's 100 smart cities. *Transactions of the Institute of British Geographers*, 43(3), 405–419.

Dowling, R., McGuirk, P. & Maalsen, S. (2018). Realising smart cities: Partnerships and economic development in the emergence and practices of smart in Newcastle, Australia. In A. Karvonen, F. Cugurullo & F. Caprotti (Eds.), *Inside Smart Cities: Place, Politics and Urban Innovation* (pp. 15–29). New York, NY: Routledge.

Elwood, S. & Leszczynski, A. (2018). Feminist digital geographies. *Gender, Place and Culture*, 25(5), 629–644.

Evans, J. (2016). Trials and tribulations: Problematizing the city through/as urban experimentation. *Geography Compass*, 10(10), 429–443.

Evans, J. P., Jones, R., Karvonen, A., Millard, L. & Wendler, J. (2015). Living labs and co-production: University campuses as platforms for sustainability science. *Current Opinion in Environmental Sustainability*, 16(1), 1–6.

Fischer, F. (2000). *Citizens, Experts, and the Environment: The Politics of Local Knowledge*. Durham and London: Duke University Press.

Goonawardene, N., Lee, P., Tan, H. W. X., Valera, A. & Tan, H. P. (2018, 2017). Technologies for ageing-in-place: The Singapore context. In T. Menkhoff, S. N. Kan, H. D. Evers & Y. W. Chay (Eds.), *Living in Smart Cities: Innovation and Sustainability* (pp. 147–174). Singapore: World Scientific.

Grossi, G. & Daniela, P. (2017). Smart cities: Utopia or neoliberal ideology? *International Journal of Urban Policy and Planning*, 69, 79–85.

Halphern, O. (2014). *Beautiful Data: A History of Vision and Reason Since 1945*. Durham: Duke University Press.

Hollands, R. (2008). Will the real smart city please stand up? *City*, 12(3), 303–320.

Howe, L. & Fox, M. S. (2017). Increasing the effectiveness of the non-profit sector through visualization: A case study of furniture banks. In K. Coperich, E. Cudley & H. Nembhard (Eds.), *Proceedings of the 2017 Institute of Industrial and Systems Engineering Conference*.

Hussain, A., Wenbi, R., da Silva, A. L., Nadher, M. & Mudhish, M. (2015). Health and emergency-care platform for the elderly and disabled people in the smart city. *Journal of Systems and Software*, 110, 253–263.

Karvonen, A., Cugurullo, F. & Caprotti, F. (2018). *Inside Smart Cities: Place, Politics and Urban Innovation*. New York, NY: Routledge.

Kim, C. (2010). Place promotion and symbolic characterization of new Songdo City, South Korea. *Cities*, 27(1), 13–19.

Kim, P., Alfaro, K. & Miller, L. A. (2015). Ecosystemic innovation for indigenous people in Latin America. *Indigenous People and Mobile Technologies*, 31, 59.

Kitchin, R. (2014). Big data, new epistemologies and paradigm shifts. *Big Data & Society*, 1(1), 2053951714528481.

Kitchin, R. (2015). Making sense of smart cities: Addressing present shortcomings. *Cambridge Journal of Regions, Economy and Society*, 8(1), 131–136.

Kitchin, R. (2019). Towards a genuinely humanizing smart urbanism. In P. Cardullo, C. Di Feliciantonio, and R. Kitchin (Eds.), *The Right to the Smart City* (pp. 193–204). Leeds: Emerald Publishing Limited. https://doi.org/10.1108/978-1-78769-139-120191014.

Kitchin, R. & Kneale, J. (2001). Science fiction or future fact? Exploring imaginative geographies of the new millennium. *Progress in Human Geography*, 25(1), 19–35.

Kong, L. & Woods, O. (2018a). The ideological alignment of smart urbanism in Singapore: Critical reflections on a political paradox. *Urban Studies*, 55(4), 679–701.

Kong, L. & Woods, O. (2018b). Smart eldercare in Singapore: Negotiating agency and apathy at the margins. *Journal of Aging Studies*, 47, 1–9.

Kourtit, K., Nijkamp, P. & Steenbruggen, J. (2017). The significance of digital data systems for smart city policy. *Socio-Economic Planning Sciences*, 58, 13–21.

Lally, N. & Burns, R. (2017, November). Toward a geographical software studies. *Computational Culture: A Journal of Software Studies* 6. http://computationalculture.net/special-section-editorial-geographies-of-software/.

Leszczynski, A. (2016). Speculative futures: Cities, data, and governance beyond smart urbanism. *Environment and Planning A: Economy and Space*, 48(9), 1691–1708.

Lin, Y., Zhang, X. & Geertman, S. (2015). Toward smart governance and social sustainability for Chinese migrant communities. *Journal of Cleaner Production*, 107(C), 389–399.

Luque-Ayala, A. & Marvin, S. (2015). Developing a critical understanding of smart urbanism? *Urban Studies*, 52(12), 2105–2116.

Lynch, C. R. (2019). Contesting digital futures: Urban politics, alternative economies, and the movement for technological sovereignty in Barcelona. *Antipode*, 52(3), 660–680.

March, H. (2018). The smart city and other ICT-led techno-imaginaries: Any room for dialogue with degrowth? *Journal of Cleaner Production*, 197, 1694–1703.

March, H. & Ribera-Fumaz, R. (2016). Smart contradictions: The politics of making Barcelona a self-sufficient city. *European Urban and Regional Studies*, 23(4), 816–830.

Marvin, S., Luque-Ayala, A. & McFarlane, C. (2015). *Smart Urbanism: Utopian Vision or False Dawn?* New York, NY: Routledge.

Moser, I. (2013). *Ageing, Technology and Home Care: New Actors, New Responsibilities.* Paris: Presses des Mines via OpenEdition.

Neff, G., Tanweer, A., Fiore-Gartland, B. & Osburn, L. (2017). Critique and contribute: A practice-based framework for improving critical data studies and data science. *Big data*, 5(2), 85–97.

Neirotti, P., Marco, A. D., Cagliano, A. C., Mangano, G. & Scorrano, F. (2014). Current trends in Smart City initiatives: Some stylised facts. *Cities*, 38, 25–36.

O'Sullivan, D. (2019). Cartography and geographic information systems. In J. Ash, R. Kitchin & A. Leszczynski (Eds.), *Digital Geographies*. Los Angeles, CA: Sage Publications: Chapter 11.

Ordendaal, O. (2018). Smart innovation at the margins: Learning from Cape Town and Kibera. In A. Karvonen, F. Cugurullo & F. Caprotti (Eds.), *Inside Smart Cities: Place, Politics and Urban Innovation* (pp. 243–257). New York, NY: Routledge.

Pollio, A. (2016). Technologies of austerity urbanism: The "smart city" agenda in Italy (2011–2013). *Urban Geography*, 37(4), 514–534.

Ramaswami, A., Russell, A. G., Culligan, P. J., Sharma, K. R. & Kumar, E. (2016). Meta-principles for developing smart, sustainable, and healthy cities. *Science*, 352(6288), 940–943.

Rosu, D., Aleman, D. M., Beck, J. C., Chignell, M., Consens, M., Fox, M. S. & Sanner, S. (2017). Knowledge-based provision of goods and services for people with social needs: Towards a virtual marketplace. In Paper presented at: *Proceedings of the Workshops at the 31st AAAI Conference on Artificial Intelligence*.

Roy, J (2001). Rethinking communities: Aligning technology & governance. In *LAC Carling Government's Review, Special Edition 6–11, June*.

Schui, F. (2014). *Austerity: The Great Failure.* New Haven: Yale University Press.

Shelton, T., Zook, M. & Wiig, A. (2015). The 'actually existing smart city.' *Cambridge Journal of Regions Economy and Society*, 8(1), 13–25.

Shin, H. B. (2016). Envisioned by the state: Entrepreneurial urbanism and the making of Songdo city, South Korea. In *Mega-Urbanization in the Global South* (pp. 95–112). New York, NY: Routledge.

Shore, C. & Wright, S. (2011). Conceptualising policy: Technologies of governance and the politics of visibility. *Policy Worlds: Anthropology and the Analysis of Contemporary Power*, 2011, 1–26.

Shwayri, S. T. (2013). A model Korean ubiquitous eco-city? The politics of making Songdo. *Journal of Urban Technology*, 20(1), 39–55.

Sterling, B. (2018). Stop saying 'Smart Cities': Digital stardust won't magically make future cities more affordable or resilient. *The Atlantic*. https://www.theatlantic.com/technology/archive/2018/02/stupid-cities/553052/?utm_source=twb

Tonkiss, F. (2018). *Urban Inequalities: Divided Cities in the Twenty-first Century*. Cambridge: Polity Press.

Trencher, G. & Karvonen, A. (2018). Innovating for an ageing society: Insights from two Japanese smart cities. In A. Karvonen, F. Cugurullo & F. Caprotti (Eds.), *Inside Smart Cities: Place, Politics and Urban Innovation* (Vol. 2018, pp. 258–274). New York, NY: Routledge.

Van Dijk, J. A. (2006). Digital divide research, achievements and shortcomings. *Poetics*, 34(4–5), 221–235.

Vanolo, A. (2014). Smartmentality: The smart city as disciplinary strategy. *Urban Studies*, 51(5), 883–898.

Veltz, M., Rutherford, J. & Picon, A. (2018). Smart urbanism and the visibility and reconfiguration of infrastructure and public action in the French cities of Issy-les-Moulineaux and nice. In A. Karvonen, F. Cugurullo & F. Caprotti (Eds.), *Inside Smart Cities: Place, Politics and Urban Innovation* (Vol. 2018, pp. 133–148). New York, NY: Routledge.

Wiig, A. (2015). IBM's smart city as techno-utopian policy mobility. *City*, 19(2–3), 258–273.

Woods, O. (2020). Subverting the logics of "smartness" in Singapore: Smart eldercare and parallel regimes of sustainability. *Sustainable Cities and Society*, 53, 101940.

Young, J. (2019). Knowledge politics. In Ash, J., Kitchin, R. & Leszczynski, A. (Eds.) *Digital Geographies*. Los Angeles, CA: Sage Publications, pp. 270–280.

https://doi.org/10.1142/9789811293108_0012

Chapter 12

Smart Cities: A Review of Managerial Challenges and a Framework for Future Research*

Thomas Menkhoff

1. Introduction

1.1. *Challenges Confronting the Future of Cities*

Cities around the world continue to be economic powerhouses and magnets for migrants from rural and suburban areas hoping to better their lives. According to UN projections, 68% of the world's population will be *urban* by 2050 (compared to 30% in 1950). Due to rapid population growth and urbanization, 2.5 billion people will be added to the world's urban population by 2050. About 90% of

---

*This is a reprint of an article by Thomas Menkhoff originally published in the *Handbook on the Business of Sustainability — The Organization, Implementation, and Practice of Sustainable Growth* (edited by Gerard George, Martine R. Haas, Havovi Joshi, Anita M. McGahan, and Paul Tracey). The editors gratefully acknowledge permission from the publisher (Edward Elgar Publishing) to reprint the article in this edition.

this increase will be concentrated in Asia and Africa (UN, 2014). There are issues that keep city leaders and managers in New York, London, Tokyo, Paris, Shanghai, Hong Kong, Singapore, Seoul, New Delhi, Jakarta, Manila, Lagos etc. up at night: poor infrastructure, unliveable housing conditions, increasing spatial density, loss of heritage, growing resource needs and a myriad of other problems ranging from traffic congestion and pollution to urban conflicts and climate change.

According to research conducted by the Lee Kuan Yew Centre for Innovative Cities, Singapore University of Technology and Design (SUTD), the biggest challenges confronting the future of cities include (i) environmental threats such as flooding, tropical cyclones (to which coastal cities are particularly vulnerable), heat waves and epidemics; (ii) difficulties to provide resources (water, food and energy) to an ever-growing urban population; (iii) social inequality as result of the widening gap between the urban super-rich and urban poor; (iv) the 'smart' use of technology (e.g., smart mobility technology) to plan, develop and run 'cities of the future' without allowing 'the emergence of a new form of social divide rooted in the technological'; and (v) 'good' governance:

"Future cities offer immense possibilities to enrich the lives of their inhabitants even as the challenges are stark. To make the best out of inevitable urbanization, good governance is imperative. Cities will increase in size and their populations become more diverse. Governing these cities will, therefore, be progressively complex and require the most dedicated of minds" (Chan & Neo, 2018).

Can liveable smart cities with their digital infrastructures provide solutions to the challenges confronting the future of cities?

1.2. *What Makes a City 'Smart'?*

Over the past two decades, the design and construction of 'smart' and increasingly 'wired' cities (Martin, 1977; Townsend, 2013) has gathered pace. 'Famous' cities with smart city features include Songdo

International Business District (South Korea) (Arbes & Bethea, 2014), featuring various smart urban infrastructure solutions related to real estate, utilities, transportation, education, health and government developed by Cisco Systems; Barcelona (renowned for its open data portal, air quality sensor network and public wi-fi in streets); and London's 'Smarter London Together' roadmap (Mayor of London, 2018).

> **Smart City (SC) Defined:** A smart city is "a developed urban area that creates sustainable economic development and high quality of life by excelling in multiple key areas: economy, mobility, environment, people, living, and government" (http://www.businessdictionary.com/definition/smart-city.html).

An ambitious transformation case is Singapore's 'Smart Nation' programme (https://www.smartnation.gov.sg/) aimed at creating competitive economy, a sustainable environment, and a high quality of life (Tan *et al.*, 2012; Khoo, 2016; Ng, 2019). With its 'Smart Nation and Digital Government Office', the Singapore Government places people at the centre of its (digital) smart nation initiatives: "We envision a Smart Nation that is a leading economy powered by digital innovation, and a world-class city with a Government that gives our citizens the best home possible and responds to their different and changing needs" (https://www.smartnation.gov.sg/why-Smart-Nation/transforming-singapore).

Concrete initiatives (Smart Nation Singapore, n.d.; see Table 3) include 'strategic national projects' such as e-payments to provide an open, accessible and interoperable national e-payments infrastructure; 'urban living' (e.g., development of a smart elderly alert system); 'transport' (e.g., development of standards for self-driving vehicles); 'health' (e.g., TeleHealth — online medical consultations); 'digital government services' (e.g., LifeSG, an easy-to-use app that helps citizens to navigate digital government services); and measures to support 'startups and businesses' (e.g., a FinTech Regulatory Sandbox in support of FinTech innovation experiments).

A basic premise of the smart city discourse is the notion that 'smart' cities can and should use digital technologies (ICT) to become more intelligent (Batty, 2012) and efficient in the use of resources with more efficient services for citizens, a higher quality of life and less impact on the environment:

> "With this vision in mind, the European Union is investing in ICT research and innovation and developing policies to improve the quality of life of citizens and make cities more sustainable in view of Europe's 20-20-20 targets. The smart city concept goes beyond the use of ICT for better resource use and less emissions. It means smarter urban transport networks, upgraded water supply and waste disposal facilities, and more efficient ways to light and heat buildings. And it also encompasses a more interactive and responsive city administration, safer public spaces and meeting the needs of an ageing population" (European Commission, 2013).

One programmatic effort of the EC's Smart City agenda is the *European Innovation Partnership on Smart Cities and Communities* aimed at a stronger collaboration between cities, industry and citizens 'to improve urban life through more sustainable integrated solutions' with regard to 'applied innovation, better planning, a more participatory approach, higher energy efficiency, better transport solutions, intelligent use of Information and Communication Technologies (ICT), etc.' (http://ec.europa.eu/eip/smartcities/index_en.htm).

In Asia, the ASEAN Smart Cities Network (ASCN), a network of 26 cities established in 2018 to foster cooperation on smart city development, also promotes *the use of smart technologies as a method of city management* amongst city leaders and urban planners in order to tackle urban problems such as reducing the urban environmental footprint via a 'sustainable built environment', 'zero carbon strategies', 'intelligent mobility systems', 'renewable energy supply' and 'Big Data analytics'.

1.3. *Smart City Components*

Despite the popularity of the Smart City discourse, there is no universally accepted definition of a 'Smart City'. Technology firms (IBM, 2012), consultants, city governments, academics and activists all use different frameworks and catchwords — albeit with a common core comprising the importance of governing, civic engagement or technology-enabled, people-centric transport solutions (e.g., in form of real-time data on bus arrival timings).

ITU's Definition of a Smart City: "A smart sustainable city is an innovative city that uses information and communication technologies (ICTs) and other means to improve quality of life, efficiency of urban operation and services, and competitiveness, while ensuring that it meets the needs of present and future generations with respect to economic, social, environmental as well as cultural aspects" (Recommendation International Telecommunication Union/ITU-T Y.4900).

Source: https://www.itu.int/en/ITU-T/ssc/united/Pages/default.aspx

---

The team that developed the 'European Smart City Model' (http://www.smart-cities.eu/model.html) defines a Smart City as 'a city well performing in a forward-looking way' with regard to six 'characteristics': Smart Mobility, Smart People, Smart Living, Smart Environment, Smart Economy, and Smart Governance (EUR-SCM, 2016; see Table 1). There seems to be widespread agreement amongst urban planners, politicians, administrators and decision makers that translating these SC components into concrete urban planning efforts will be instrumental in ameliorating urban problems and creating smart(er) cities (Cohen & Muñoz, 2016; Apte, 2017; http://www.smart-cities.eu/download/smart_cities_final_report.pdf).

Smart Governance puts emphasis on strong SC leadership and regulatory environments. Other factors include the provision of affordable and needs-based public and social services; participation of citizens in decision-making processes; strategic SC continuity in case the reins of government are handed over to an opposition party; and forward-looking strategic SC planning and integrated master planning approaches. According to the 'Smart City Mission' of the Government of India, core urban infrastructure elements include: "(i) adequate water supply, (ii) assured electricity supply, (iii) sanitation, including solid waste management, (iv) efficient

Table 1.   Characteristics and Factors of a Smart City

Smart Economy	Smart People
• Innovative spirit	• Level of qualification
• Entrepreneurship	• Affinity to lifelong learning
• Economic image & trademarks	• Social and ethnic plurality
• Productivity	• Flexibility
• Flexibility of labour market	• Creativity
• International embeddedness	• Cosmopolitanism / Open-mindedness
• Ability to transform	• Participation in public life
Smart Governance (Participation)	**Smart Mobility**
• Participation in decision-making	• Local accessibility
• Public and social services	• (Inter-)national accessibility
• Transparent governance	• Availability of ICT-infrastructure
• Political strategies and perspectives	• Sustainable, innovative and safe transport systems
Smart Environment **(Natural resources)**	**Smart Living**
• Attractivity of natural conditions	• Cultural facilities
• Pollution	• Health conditions
• Environmental protection	• Individual safety
• Sustainable resource management	• Housing quality
	• Education facilities
	• Touristic attractivity
	• Social cohesion

Source: http://www.smart-cities.eu/download/smart_cities_final_report.pdf, p. 12.

urban mobility and public transport, (v) affordable housing, especially for the poor, (vi) robust IT connectivity and digitalization, (vii) good governance, especially e-Governance and citizen participation, (viii) sustainable environment, (ix) safety and security of citizens, particularly women, children and the elderly, and (x) health and education" (http://smartcities.gov.in/cityChallenges.aspx).

Affinity to life-long learning is one of the *Smart People* sub-components. In Singapore, a key driver behind inculcating the importance of life-long learning into students and the workforce is *SkillsFuture Singapore* (SSG), a statutory board under the Ministry of Education (MOE): "It drives and coordinates the implementation of the national SkillsFuture movement, promotes a culture and holistic system of lifelong learning through the pursuit of skills mastery, and strengthens the ecosystem of quality education and training in Singapore" (https://www.ssg-wsg.gov.sg/about.html). Strategic goals include personalised skills upgrading, developing an integrated, high-quality system of education and training that responds to constantly evolving industry needs and meeting the demands of different sectors of the economy. Other people components are creativity, cosmopolitanism, open-mindedness and participation in public life (see Table 4). To create smarter cities, 'self-decisive', 'independent' and 'aware' citizens with a 'civic sense' are seen as indispensable (http://www.smart-cities.eu/download/smart_cities_final_report.pdf).

Smart Mobility refers to safe and effective road transportation embedded in the city's infrastructure that includes cars, bicycles, buses, trains, walking, road and expressway designs as well as road pricing methods to regulate traffic congestion. An example of a technology-enabled road pricing system is Singapore's Electronic Road Pricing (ERP) and its touted successor ERP-2, which are enhanced road toll collection methods intended to avoid gridlocked traffic. Such computerized implementations do require careful policy and strategy considerations by traffic authorities prior to implementation, e.g., in order to decide whether to charge by per-entry when a car enters a pre-defined busy area, or to charge by distance travelled from point A to point B (Phang, 2018). Other components comprise the promotion of cycling and the creation of a walkable city.

Smart Living components emphasize access to quality housing, social cohesion, health, safety and education. Social cohesion can be measured based on the extent to which a community of residents sharing common city space as fellow dwellers is indeed tightly knitted. A related question is whether there is a common city identity that most people are proud of. Suitable health conditions imply having access to affordable healthcare. Individual safety is a key element aimed at ensuring a low crime rate, limiting access to the tools of crime, and an effective police force. Other sub-components are cultural facilities such as museums and libraries as well as touristic attractivity.

An amenable *Smart Environment* with a good "balance between built-up space and green areas, water resources, pollution control and use of resources in a responsible and environment friendly way (for example, use of renewable energy, rain-water harvesting, green initiatives and so on)" (Apte, 2017) is an important component of a smart city. Pollution affects people's quality of life in cities. Increasingly, sensor network technology is used to monitor pollution levels (Luo & Yang, 2019). Other urgent concerns include the '5R's': refuse, reduce, reuse, repurpose, and recycle as indicated by the low recycling rates around the globe as documented by The Waste Atlas developed by the University of Leeds (http://www.atlas.d-waste.com/). Related factors comprise attractivity of natural conditions and sustainable resource management (see Table 3).

Trendsetting is Singapore's Deep Tunnel Sewerage System (DTSS). As a 'used water superhighway for the future', DTSS represents a sustainable solution which was conceived by Singapore's National Water Authority (PUB) to meet Singapore's long-term needs for used water collection, treatment, reclamation and disposal (Rahman, 2018).

A *Smart Economy* requires a strong ecosystem of entrepreneurship and innovation, i.e., the existence of entrepreneurial leaders, effective ideation and innovation management, access to venture capital funding, go-to-market know how etc. Cities which are unsafe and culturally unattractive will find it difficult to attract enough innovation talent in order to excel. An optimal economy is driven by innovation and the allocation of resources in the right direction. Unlike other small countries whose economies are heavily dependent on just one or two sectors, Singapore started its diversification programmes early based on a cluster-based development approach. During the past few years, a new start-up cluster has been successfully established, and efforts are underway to build a new A.I. ecosystem (https://www.aisingapore.org/).

1.4. *Critique of the Smart City Concept*

The smart city is, to many critical observers, just a 'buzzphrase' that has been envisioned and promulgated by big technology, engineering and consulting firms 'all of whom hoped to profit from big municipal contracts' (Poole, 2014). Sennett (2012) regards the smart city as 'over-zoned, defying the fact that real development in cities is often haphazard, or in between the cracks of what's allowed'. Shelton *et al.* (2015, p. 13) have called the smart city concept "a somewhat nebulous idea, which seeks to apply the massive amounts of digital data collected about society as a means to rationalise the planning and management of cities".

Many cities find it challenging to finance grand urban connectivity projects and to ensure that master plans are well designed and executed. In many Asian countries, material infrastructure development is more important than digital development via SC technology-enabled road pricing systems, smart transport apps or smart(er) streetlamps with facial recognition technology. The hype around 'digital urban solutions' can easily distract from urgently required material improvements and the need to tackle social inequalities.

As Zelton *et al.* (2015, p. 21) have stressed, "In Philadelphia, the smart city has acted primarily as a promotional vehicle, highlighting the city's efforts to produce a competitive, entry-level workforce for the 21st century economy, despite achieving few meaningful results in this respect". Other smart city cynics (e.g., Greenfield, 2013) have argued that the idea of a top-down, turnkey smart city isn't more than a techno-utopian fantasy propagated by big and powerful corporations.

There are also increasing privacy concerns over SC technologies such as 'city surveillance solutions' (Mitchell & Heynen, 2009) and unintended effects of crowd analytics. There are approximately 5.9 million CCTV cameras in the United Kingdom where an average person would be captured by 70 CCTV cameras on a normal day (Temperton, 2015).

To what extent smart digital technology 'solutions' can indeed resolve problems caused by rapid urbanization in developing Asia, is a poorly researched topic (Lee *et al.*, 2020). According to social scientists who have studied the impact of technologies in smart cities on their inhabitants' lives, there are many 'antecedent challenges' such as unanticipated user needs, power, resistance to power, and inequality which can thwart visionary smart city (digital) development initiatives and projected habituation rates of greenfield cities (Shepard, 2017; Ng, 2020).

Other challenges arise from difficulties of making real 'inclusion' work or to enable SC executives to 'think in data' to make digital visions and plans actionable. To develop and effectively organise the smart IT systems underlying smart cities is a challenging and very complex task for governing bodies, SC leaders and managers as the 'Job Description of a Smart City Specialist' in Appendix 1 shows.

Let us now take a closer look at some typical SC applications in order to appreciate the manifold managerial challenges 'SC specialists' are facing who have been tasked to make a smart city work.

2. Smart City Applications

By default, SC applications are designed to help SC stakeholders such as urban commuters 'to get a job done', e.g., related to individual *Smart Mobility* (see Table 2). One example is Parking.sg, a mobile application developed by the Government Technology Agency of Singapore ('We harness the best info-communications technologies to make a difference to the everyday lives of people in Singapore'). Through the mobile app, users can pay short-term parking fees at coupon-based public car parks. The app serves as an alternative mode of payment to paper parking coupons. Users who have turned on the 'Notification' feature in their device settings, can be alerted 10 minutes before their parking session expires. They can also extend their parking duration remotely via the app (https://www.tech.gov.sg/).

IoT-enabled Smart Parking services are related applications that help users to find free parking spots in the city. The IoT-based sensing system (using sensors and microcontrollers located in carparks) sends data about vacant spaces for parking via a web/mobile application to the driver which reduces search traffic on streets. Table 2 provides a brief overview about the wide spectrum of SC applications (Novotný *et al.*, 2014).

Under the category 'Open Data & Analytics for Urban Transportation', joint R&D efforts are underway by both public and

Table 2.   Smart City Components (and Applications)

Smart Environment	Smart Living	Smart Governance
(smart grids / smart metres cutting energy costs and more accurate bills)	(operationally and secure, efficient smart city buildings)	(faster, more productive and more economical public services)
Smart Mobility	**Smart People**	**Smart Economy**
(real-time location aware services that meet the spatial information needs of people in order to reduce travel time and avoid traffic delays)	(a cloud-based smart mobility platform offering data-driven and needs-based shuttle bus services)	(municipal e-services that help residents enter the labour market / SGFinDex that relies on consent and a 'national digital identity')

private sector organisations as well as startups to support people by giving them visual and tangible access to *real-time information* about their own city so that they can make *better decisions* about healthier routes to work, convenient (empty) parking lots, attractive dining options or the nearest bicycle for rent (https://www.smartnation. gov.sg/what-is-smart-nation/initiatives/Transport).

An example of a *Smart Living* application is a smart-enabled home for the elderly as piloted by the TCS-SMU iCity Lab and the so-called SHINESeniors project with focus on elderly Singapore citizens living alone in Singapore's Housing Development Board (HDB) rental flats. Each flat is covered by several passive infrared (PIR) sensors. In addition to the electro-magnetic reed switch which is attached to flat's door, each home is equipped with a gateway that relays sensor data to the back-end for storage, analysis, and visualisation as a means of unobtrusive (actionable) in-home monitoring (Goonawardene *et al.*, 2018).

Environmental protection, reduction of water losses, climate change mitigation by replacing fossil-based materials with more renewable ones aimed at reducing CO_2 emissions represent other application goals that can be achieved with the help of sensor networks.

Table 3. Examples of Singapore's Smart Nation Programmes*

Focus Area	Strategic Goals
Smart Digital Governance: **National Digital Identity**	Allowing individuals to prove their legal identity digitally via SingPass Mobile app, e.g., signing documents and contracts easily and securely (removes the need for physical presence and paper-based signing).
Smart Digital Economy: **E-Payments**	Providing seamless, secure, and integrated e-payment platforms, options for cashless payments, and integrating e-payments into business processes from end to end (regulated by the 2020 Payment Services Act). It is planned to phase out cheques from 2025 onwards.
Smart People/ Digital Society: **LifeSG**	Providing people access to technology so that they can effectively connect with government services with just one app (e.g., childbirth registration): LifeSG.

Note. *More information is available at: https://www.smartnation.gov.sg/.

Table 4. Data-Driven Nudging People to Adopt Eco-friendly Transport Modes (Walking) in Support of SC Components

SC Components	Behavioural Aspects
Environment	Use of eco-friendly transport modes: bicycles, e-bikes, e-vehicles, greener trains, e-motorcycles, multiple occupant vehicles, hybrid vehicles, pedestrians
Governance	Engaging citizens in a nudging pilot project 'to make them walk' (more) with the help of message-driven nudging
People	Making (nudging) people (to) walk in support of community health, vitality and safety
Living	Availability of safe spaces and interesting places/routes for people to walk
Mobility	Easy access (within walking distance) to destinations
Economy	Attractive streetscapes for pedestrians with shopping opportunities

The IoT connectivity revolution has also impacted digital supply chain management, e.g., by using predictive analytics to optimise inventory allocation and forecast demand. Sensor technology is increasingly used to improve the quality of shipment conditions (by monitoring vibrations, strokes, container openings or cold chain maintenance for insurance purposes), item location (search of individual items in big surfaces like warehouses or harbours), storage incompatibility detection (warning emission on containers storing inflammable goods close to others containing explosive material) and fleet tracking (Libelium, 2020).

Advances in new digital SC technologies is generating a huge amount of data and information with vast new business opportunities, e.g., in the area of big data analytics. Data mining can help to infer attitudes and preferred lifestyle choices of citizens through sentiment analyses of microblogs (Hoang *et al.*, 2013) or the analysis of transportation-related pain points via behavioural insights deduced from large-scale taxi trip data. Understanding the behavioural profiles of SC citizens in 'smart' home environments derived from home-embedded sensors and biometric wearables can be a valuable source of analytical insights. But there are also issues such as data security, privacy concerns, inadequate data systems governance or if consumers don't trust remote monitoring devices.

Hi-tech or low-tech induced *nudging* (Thaler & Sunstein, 2008) qua text messages or letters represents an interesting SC use case to highlight both the new opportunities of digital SC applications *enabling real-time, smart decisions* about living healthily, using public transport, participating in public life or to save water and energy as well as associated issues. On the one hand, data-driven nudges can help cities to achieve their sustainability goals and promote eco-friendly transport modes by connecting traffic sensors to messaging systems so that commuters take public transport at times of high congestion. On the other hand, message-driven nudging in datafied cities raises legal-ethical issues because of concerns that nudgers employ tricks to get us to do what *they* want or that nudging erodes people's responsibility for their own choices (Schmidt & Engelen, 2020). SC applications such as well-intended, analytical nudging efforts raise several *managerial challenges* for SC stakeholders be it city managers, SC businesses or citizens such as concerns about unacceptable paternalism or cost.

3. Smart City Managerial Challenges

In the following, we shall present a couple of examples to explain how SC-specific applications related to the six SC components raise some managerial challenges. The first example is related to the SC category 'Smart Governance' (Transparency and Collaboration) and underscores the criticality of involving multiple stakeholders in policy development and implementation. A recent case in point related to Singapore's national FinTech strategy is the development of a *data-sharing platform* that can train models to improve credit assessments (in support of SME financing and post-pandemic recovery) as announced by Mr Heng Swee Keat, Singapore's Deputy Prime Minister and Coordinating Minister for Economic Policies / Minister for Finance, at the 12/2020 Singapore FinTech Festival x Singapore Week of Innovation & TeCHnology (SFF X SWITCH). Stakeholders include the Monetary Authority of Singapore (MAS), the National Research Foundation (NRF), National University of Singapore (NUS), lenders and small businesses (MAS, 2020). A related SC breakthrough innovation is the Singapore Financial Data Exchange System 'SGFinDex', "the world's first public digital infrastructure to use a national digital identity and centrally managed online consent system to enable individuals to access, through applications, their

Table 5. Smart City Managerial Challenges in Relation to SC Components

Governance	Individual privacy and secure, ethical and efficient data sharing/ strong public private partnerships (PPPs)
People	Participative placemaking/incentive alignment/strong PPPs/ active citizenship
Living	Preventing siegeware attacks on home & building owners/ cybersecurity matters
Environment	Sustainable public-science collaboration and participative decision-making/PPPs
Economy	Value creation through the use of corporate 'city business models'
Mobility	Achieving real-time, 'smart' mobility in a multi-stakeholder ecosystem/managing stakeholder plurality

financial information held across different government agencies and financial institutions" (https://www.mas.gov.sg/development/fintech/sgfindex).

An important managerial challenge (besides developing the institutional, physical, socio-economic and ICT infrastructures) is to effectively *govern* such data-exchange platforms and to preserve data privacy as evidenced by Singapore's voluminous *2019 Trusted Data Sharing Framework* developed by the Infocomm Media Development Authority of Singapore and the Personal Data Protection Commission (IMDA/PDPC, 2019).

Potential research questions: What are the ethical-legal concerns arising from collecting and processing large amounts of personal and impersonal data used to influence citizens' behaviour? How to effectively manage the privacy and trust implications that arise from the need to comply with data protection laws, consent frameworks, data portability rights? How to embed trust in SC platform-related engineering processes? How can less developed nations catch-up with the FinTech governance approaches adopted by globally leading fintech hubs?

The second example relates to the SC category 'Smart People' (Well-Being of People). Relevant SC components include 'participation in public life', 'citizen engagement' and 'co-creation' (e.g., place design). Specific SC applications comprise 'participatory placemaking' (Pak, 2018) and a 'community-centric, collaborative design process'. The Placemaking Europe Network defines participatory placemaking as follows: "Turning spaces into places that increase the presence of people in public spaces through the participation of users, the collaboration of stakeholders and by signalling shared ownership of the common urban spheres" (URBACT, 2019). Public spaces include large inner-city public squares, parks, beaches, streets and other urban natural environments such as woodlands or riverbanks. Participatory placemaking requires a high level of active citizenship and strong partnerships with relevant stakeholders aimed at "sustaining public spaces as urban commons and creating pacts of joint responsibility for developing and maintaining such spaces" (URBACT, 2019).

URBACT is a European exchange and learning programme promoting sustainable urban development, comprising 550 cities, 30 countries and 7,000 active local stakeholders. It is jointly financed by the European Union (European Regional Development Fund) and the Member States. URBACT develops and shares new and sustainable solutions to major urban challenges, good practices and lessons learned with urban policy stakeholders, integrating economic, social and environmental (ESG) dimensions (https://urbact.eu/urbact-glance).

A managerial challenge of participatory placemaking is the competency of leaders and planners to effectively *support* communities and local active citizens in line with the guiding principles for good placemaking projects such as evidence-based placemaking and 'to acknowledge the *agency of citizens* to make changes and improvements' (URBACT, 2019). This arguably necessitates (more) empathy for the 'Right to Cities' movement (whose goals are in conflict with the basic principles of private ownership and profit generation which regulate many urban spaces today) and acknowledging the valuable role of the citizenry in shaping (equitable) urban habitats. Events such as the eviction of homeless people from public places, gated communities, gentrification or violent public place clashes such as the Taksim Gezi park uprising in Istanbul in 2013 suggest that the vision of inclusive and compassionate place-making remains elusive in many cities. Top-down city planning systems are still predominantly in place in many Asian countries. Therefore, more research is necessary to figure out how to make participatory place-making work in the diverse Asia Pacific region (with vast differences in terms of politics, language, culture and region) marked by disparities across economies and dynamic ethnocscapes (Chun, 2012).

Potential research questions: How can "improved place-making approaches" lead to greater inclusive quality of life improvements for all urban residents, "protecting and enhancing access to public space while retaining principles of affordability and accessibility" (Hoe, 2018)? This could entail to comparatively examine to what extent government-led placemaking policies and practices differ across selected Asian countries, the impact of tightly controlled urban planning and cultural policies on 'successful' urban regeneration initiatives (as in Singapore) and the feasibility to apply some of the European place-making practices with their emphasis on strong and active citizen participation in Asia's (emerging smart) cities such as Jakarta, Manila, Kuala Lumpur, Hanoi and so forth. More evidence is required that (and if yes, how) 'smart' people can indeed positively impact 'smart' cities with their complex ethnoscapes.

C40 Cities

C40 is a network of the world's megacities committed to addressing climate change. C40 supports cities to collaborate effectively, share knowledge and drive meaningful, measurable and sustainable action on climate change (https://www.c40.org/about).

Another important SC dimension is 'Smart Living' (Smart Enabled Buildings). Related SC components include Internet of Things (IoT) and sensor technologies (e.g., for intelligent city buildings), enabling core SC applications such as (i) real-time, smart buildings related intelligent decisions (based on the analysis of connected IoT devices) to ensure the secure operational efficiency of city buildings (e.g., in terms of energy conservation) and/or (ii) efficient and real-time SC security systems and high-speed communication security protocols that provides strong intrusion detection. Issues such as gaining unauthorized access to data in a computer system or eavesdropping private conversations in buildings and homes which raise security and privacy concerns point to various managerial challenges affecting top management teams tasked with developing IoT-enabled, smart buildings/homes.

A case in point is generating smart buildings related data ('Big Data') at remote locations and *transmitting* them safely to central city servers for further actionable analyses (Rathorea *et al.*, 2018). Imagine an operational manager in a property company that manages several buildings in several cities. Would a manager (with limited cybersecurity skills) be capable enough to deal with the following text message? "We have hacked all the control systems in your building at 200 Church Street and will close it down for three days if you do not pay $60,000 in Bitcoin within 24 hours." Modern building automation systems manage heating, air conditioning and ventilation, as well as fire alarms and controls, lighting, and security systems. Combining criminal intent with poorly protected remote access to software that runs building automation systems, 'siegeware' (i.e., 'code-enabled ability to make a credible extortion demand based on digitally impaired building functionality') becomes a very real possibility (Cobb, 2019).

Potential research questions: What needs to be done in terms of human capital development to reduce the risk of siegeware attacks and to cope with the fact that any prevention requires deep IT/cybersecurity competencies which might be in short supply, especially in poorer countries? To what extent do cybersecurity management approaches differ in both resource-rich and resource-poor contexts with what effects? To what extent can benchmarking efficient and real-time Smart City security systems (that are functionally cost effective, secure and able to work in a real-time, high-speed Smart City environment) help city councils to be in better control of siegeware attacks, and how will that contribute to strong(er) intrusion detection at intelligent city buildings?

A very critical aspect of the SC discourse is the achievement of a 'Smart Environment' in terms of environmental protection and sustainable resource management. An increasingly popular tool to achieve that is 'citizen science', i.e., the participation of citizens in scientific research to increase scientific knowledge, e.g., in the context of urban climate change adaptation: "Citizen science offers volunteers the opportunity to engage in environmental research while participatory modeling engages individuals in community-level environmental decision-making" (Gray *et al.*, 2017). A managerial challenge (especially in developing countries) is to master the required competencies to enable 'learning through modelling practices', e.g., with the help of participatory modelling software as postulated by Gray *et al.* (2017) in order to spur the development of 'self-organized and co-created conservation action' on the basis of participatory environmental decision-making with multiple stakeholders. But as Wamsler *et al.* (2020) have stressed there is very little empirical evidence that supports the notion that involving citizens in nature-based approaches for urban climate change adaptation helps to ensure a transformative adaptation process in cities.

Potential research questions: Where is the evidence that inclusive forms of participatory conservation planning with citizens as volunteers in terms of public-science collaboration can improve environmental decision-making and respective conservation outcomes? A related topic concerns the business model innovation approaches of start-up organizations such as Handprint, Pachama (Shopify integration), EcoMatcher, Almond or Poseidon that are enabling novel carbon offsetting services in support of a more carbon conscious society. To what extent do such digital B2B service platforms help companies to go green with positive results pertaining their sales channels and to become 'earth-positive' (https://startup.network/startups/424007.html)?

Jouliette (Amsterdam)

An "intelligent" approach to smart(er) urban sustainability is the Jouliette platform service (named after the Joule unit of measurement for energy) at the De Ceuvel social innovation community in northern Amsterdam (the Netherlands), which consists of 16 office buildings, a greenhouse, a restaurant, and a bed and breakfast — all connected to a private, behind-the-meter smart-grid. Through a new blockchain-based energy sharing token (named the 'Jouliette'), individuals and communities can manage and share their locally produced renewable energy (https://spectral.energy/news-3/jouliette-at-deceuvel/).

In terms of creating cities with a 'Smart Economy', Timeus' *et al.* (2020) have suggested that city councils should utilise the 'city business model' (based on the Business Model Canvas for firms) as practical framework to design, deliver and assess 'smart services' as well as their 'expected economic, environmental and social impacts'. Their article explains how such a planning approach was used to design a strategic ICT platform in Bristol in support of the city's four 'resilience' goals (British City Council, n.d.): (i) 'fair' (every person in Bristol has the assets and opportunities to enjoy a good life), (ii) 'liveable' (the city centre and neighbourhoods are great places for people of all ages to live, work, learn and play),(iii) 'sustainable' (The city and region prosper within environmental limits through adopting new behaviour and technology); and (iv) 'agile' (Bristol citizens and leaders make decisions based on shared priorities and real-time information).

While the use of 'smart urban business models' might be instrumental in attracting more investors to the city, a related managerial challenge is to overcome stakeholders' concerns about the relevance and value added of such 'corporate' planning methods, e.g., as in the case of the elderlies who are unfamiliar with ICT platforms. Others include financial barriers and replicating proven models of successful SC use cases elsewhere. There is a general lack of crowd-funding platforms where citizens and institutions can participate in the planning and financing of SC projects. Anecdotal evidence suggests that online platforms such as *Smartcity.brussels* which allow citizens to submit ideas, select projects and discuss complex issues are not (yet) very widespread in Asian cities (https://smartcity. brussels/the-project).

Potential research questions: In what ways have 'city business models' impacted the performance of SCs in Asian and non-Asian countries? What drives the success and failure of such platform approaches? How inclusive and value added are 'canvas driven urban ICT platforms' — for various types of stakeholders? How can successful SC use cases (e.g., crowdfunding platforms) be more effectively shared, scaled and sustained?

With regards to 'Smart Mobility', data integration across different platforms is a central component to provide real-time travel information applications that enable commuters to effectively plan trips on private and public transportation. An *effective data integration architecture design* is a must to ensure that the workflow of data collected from multiple sources (e.g., 'big' taxi trip data) creates consistent, conformed, comprehensive, clean, and current information for further real-time analyses, decision making purposes and speedier services, e.g., by matching commuter demand with driver/ bus supply.

The managerial challenge is to achieve SC goals related to 'smart' mobility in a multi-actor ecosystem increasingly influenced

---

San Francisco's On-Street Shared Vehicle Permit Program

The goal of the On-Street Shared Vehicle Permit Program (approved by the San Francisco Municipal Transportation Agency/SFMTA in 2017 after a pilot measure of about 210 on-street car share spaces at 140 locations across S.F.) is to better manage parking demand. SFMTA's overall strategic goal is to "make transit, walking, bicycling, taxi, ride sharing and carsharing the preferred means of travel". Challenges included concerns of some neighbours who didn't like on-street spaces used for this purpose; theft and vandalism of shared vehicles; implementation coordination; and construction and street closures (https://www.sfmta.com/sites/default/files/projects/2017/Carshare_eval_final.pdf).

---

by MaaS (mobility-as-a-service) and the trend to integrate public transport services into ride options apps (showing the best affordable and convenient ride types be it bike, scooter, public transit etc.), e.g., by finding common ground in view of an (over)supply of private-hire vehicles (ridesharing) and government-linked taxi businesses. The bottomline is that the data integration architecture design effectively addresses issues related to 'scalability, reliability, availability, and fault-tolerance' (Harris & Sartipi, 2019).

Potential research questions: Where is the evidence that related techniques such as geohash which alert taxi drivers (e.g., Grab) to head to hot spots where demand outstrips supply actually motivate drivers to go to those spots once the system has notified them? In what ways does the MaaS service concept (that aims to integrate public transport with other mobility services, such as car sharing, ride sourcing, and bicycle sharing) make it easier for users to plan, book, and pay for complementary mobility services, thereby facilitating less car-centric lifestyles?

Copenhagen's 'GreenWave'

Copenhagen has a bicycle-commuting rate of about 40%. Through the establishment of the 'Greenwave' system that gives priority signals to bicycles, the city managed to support eco-friendly mobility and to reduce carbon dioxide emissions by more than 90,000 tons per year (https://www.centreforpublicimpact.org/case-study/green-waves-bicycles-copenhagen/).

---

4. Theoretical Frameworks for Smart City Research

There are many questions about the benefits and drawbacks of SCs which remain unanswered to date. Some have become research subjects in SC-related areas ranging from management to information systems. Others are still poorly researched, and important knowledge gaps remain such as the risks of using A.I. in SC applications (as indicated by the potentially discriminatory effects of A.I.). A closer look at existing SC studies suggests that more conceptual-empirical research on the following issues is necessary to arrive at theory-driven answers that can improve our understanding of the broader SC phenomenon. Such works must eventually support SC leaders, managers and other SC specialists tasked to create more liveable and sustainable cities.

The broad scope of the potential research questions outlined in Sections 3 and 4 of this paper points to the fact that our knowledge about the antecedents, functioning and consequences of SCs is still rather limited.

In the following, we shall evaluate several theories that we think are relevant to better understand SC issues in order to demonstrate the powerful insights that SC stakeholders tasked with the development and management of SCs can gain by utilizing postulations and ideas that explain and interpret facts. We shall focus on five theories which we believe are critical for a better understanding of SC matters: (i) critical urban theory; (ii) theories of governance; (iii) behavioural science theories; (iv) theories of networks and ecosystems; and (v) theories of technology adoption and business model innovation (see Table 6).

4.1. Critical Urban Theory

There is a wide variety of *theories* urban leaders and managers should be exposed to in order to 'manage' urbanization and SC matters. While some of these theories can be considered as 'useful', others appear to be somewhat 'impractical' at first sight (at least from a managerial point of view) because of the way they challenge and extend existing knowledge about urban phenomena. One example

Table 6. Examples of Relevant (Managerial) Theories and Broad SC Research Topics for Further Research

Relevant (Managerial) Theories	SC Research Topics
Critical Urban Theory	• Participative placemaking • Achieving 'real' digital inclusion • A.I. legislation to protect fundamental rights
Governance Theory	• 'Good' (open) data sharing governance and management frameworks • Public-science collaboration for the benefit of sustainable resource management • Effective governance of civic SC engagement (online and offline) and outputs
Behavioral Science Theory	• Influencing ('nudging') citizens'/consumers' behaviour towards more eco-friendly habits • Combining SC technologies with behavioural insights from behavioural economics, political theory and the behavioural sciences to influence ('nudge') the behaviour and decision making of groups • Impact and outcomes of message-driven nudging on citizens/consumers
Theories of Networks and Ecosystems	• Assuring secure, ethical and efficient data sharing (data commons) as basis of connected ecosystems • Providing really 'smart' mobility solutions based on effective multi-stakeholder networks (MaaS) • Design of collaborative partnerships (platforms, living labs, crowdfunding) with a commitment to open data, interoperability and integration (open APIs) aimed at innovation exchanges and better ways to address urban issues
Theories of Technology Adoption and Business Model Innovation	• Creating liveable and sustainable cities with appropriate, needs-based SC technologies • Creating and capturing new value through 'city business models' • Effective cybersecurity management

of the latter category is *critical urban theory* which offers a lens to view urbanization matters from a political economy point of view by theorizing about a more socially just and sustainable form of urbanization without inequality and exploitation.

Critical urban theory protagonists such as Harvey (1989), Brenner (2004, 2009), and Brenner & Theodore (2002) regard cities as social and material *spaces* produced and reshaped by capital interests (McGuirk, 2004). Despite the often negative tone of critical urban theory publications ('capitalism annihilates space to ensure its own reproduction') with their emphasis on capitalism-state relations, urbanization of capital (Christophers, 2011), social exclusion, lack of justice, 'Right to the City' struggles triggered by unfavourable socio-economic conditions or contested urban commons (Harvey, 2013), city leaders and managers tasked to turn SC visions into reality would arguably be well advised to acknowledge some of the key hypotheses to better understand the rapidly changing realities of everyday urban life in cities as theatres of global transformations and to adopt a more 'holistic' SC management perspective.

Valuable takeaways of 'reading critical theory again' in relation to the SC trend could be (besides intellectual stimulation in general) a better understanding of (i) the drivers and consequences of *gentrification* processes (Bernt, 2012) that are pushing low-income groups out of upscaling neighbourhoods or (ii) the emergence of new urban movements such as the 'right to the city' initiatives against profit-oriented urban policies such as the *Derecho a la Ciudad movements* in Latin America (Eizenberg, 2012; Rutland, 2013; Domaradzka, 2018). Examples such as the 2011 Occupy Wall Street movement in New York City's Zuccotti Park that claimed public space or the violent 2013 Taksim Gezi Park demonstrations in Istanbul against urban development plans (Letsch, 2013) underscore the explosive force of Harvey's (2016) urban commons perspective ('I like the idea of an urban common which is a political concept which says that this space is open for all kind of people') in view of increasingly corporatized, inner city public spaces, police raids on protesters' encampments and associated freedom of assembly restrictions.

A related present-day 'public sphere paradox' with massive implications for further research is the existence of sophisticated communication technology and the apparent lack of a culture of citizen participation (Fraser, 2014). SC critics argue that SC technologies are not serving the needs of 'people'. More and more urban commons (defined as a social practice of governing a resource such as inner-city parks) are managed by state or market actors but not as a community of users that self-governs it through institutions it has created. In many cities, the vision of urban openness has yet to be achieved.

How to achieve 'real' digital inclusion of people in an era of 'open data' and 'urban openness' so that citizens can participate by identifying and solving urban problems as well as co-creating service prototypes? How should urban leaders deal with the urban commons challenge such as social movements aimed at 'liberating' 'regulated' public space from state organisation and state activities so that it becomes 'common space for people'? How best to respond when planners tasked to redevelop obsolete infrastructure as public space become the victims of their own success as happened in the case of New York's High Line (an elevated linear park, greenway and rail trail created on a former New York Central Railroad spur on the west side of Manhattan) that spurred real estate development in adjacent neighbourhoods, increasing real-estate values and prices along the route? How to convince sceptics that community education pays in raising awareness and promoting a more sustainable way of living by utilising data-driven hi-tech solutions for a better quality of life for residents? What are some examples of successfully managing cultural and natural resources ('commons') for collective benefits such as air, water, and a habitable earth?

Digital Twins

Trend-setting in terms of 'urban openness' and 'open data' are 'digital twin' projects such as Helsinki's Kalasatama Project (a new seaside district under development in Helsinki) in which residents, businesses, architects, city modeling specialists and other stakeholders in the area collaborated to draw out solutions. The digital twin data are open for anyone to use in order to create new services based on them. 'Helsinki's Energy and Climate Atlas, for example, uses the million semantic surfaces of 80,000 buildings to calculate and visualize the city's solar energy potential' (https://aec-business.com/helsinki-is-building-a-digital-twin-of-the-city/).

4.2. Theories of Governance

An important factor that has an impact on Smart Cities is governance (see Table 7). 'Strong' governance is consensus oriented, accountable, transparent, responsive, effective and efficient as well as equitable and inclusive based on the rule of law. Australia's Governance Institute (AGI), for example, defines it as follows: "Governance encompasses the system by which an organisation is controlled and operates, and the mechanisms by which it, and its people, are held to account. Ethics, risk management, compliance and administration are all elements of governance" (https://www.governanceinstitute.com.au/resources/what-is-governance/).

Key governance sub-components include strong SC leadership in terms of the mayor's commitment towards SC development, a dedicated organization to support SC goals, SC strategy execution (e.g., based on an implementation roadmap) and SC regulations.

Table 7. Elements of Smart Governance

Transparency	A means of holding public officials accountable and fighting corruption.
Collaboration	Involves the government, community/citizens and private sector communicating with each other and working together to find solutions, e.g., for urban problems.
Participation and Partnership	Citizen participation is a cooperative arrangement between the government and communities aimed at completing a project and/or to provide services to the population.
Communication	Skilful and transparent communication between government and stakeholders enhances citizen engagement in political systems.
Accountability	According to the Australian Public Service Commission (APSC), accountability involves being called to account to some authority for one's actions: "In a democratic state, the key accountability relationships are between citizens and the holders of public office, and between elected politicians and bureaucrats".

Source: Adapted from Ferro Guimarães *et al.* (2020).

Some cities have introduced key performance indicators (KPIs) to measure SC policy impact and effectiveness of SC digital technologies with regards to urban operations and services, quality of life improvements; and better ways to cultivate environmental sustainability. According to the ITU (International Telecommunication Union) Academy, KPI dimensions include information and communication technology; environmental sustainability; productivity; quality of life; equity and social inclusion; and physical infrastructure.

Little is known about the adoption rate of international KPI standards such as 'Recommendation ITU-T Y.4903/L.1603 — Key Performance Indicators for Smart Sustainable Cities to assess the Achievement of Sustainable Development Goals' in Asian cities as part of urban governance approaches.

A key normative feature of smart(er) governance is participation in decision-making and the need of government to seek opinions of citizens before making important decisions, e.g., in the context of a national referendum. 'Smart' governance, with the help of ICT, is expected to enable the collaborative participation of various actors in SC-related decision-making processes.

Citizen Participation

The "'Brussels Hacks The Crisis' project is a citizens participation project which invited everyone in Brussels to share their innovative and digital ideas to imagine a world post Covid. More than 100 ideas were submitted on a participation platform from 4 to 21 June by enthusiastic citizens. These ideas were then examined by a jury and submitted to a vote..." (https://smartcity.brussels/brusselshacksthecrisis-en).

The quality and 'goodness' of actual governance approaches differ from city to city and country to country, i.e., some city councils may consider governance as a form of *bureaucratic* governance with closed structures which may lead to estrangement between career public servants, their political superiors, and the public, while other city management teams may emphasize inclusiveness and empowering citizens as politico-economic SC goals.

A related governance issue is the impact of more *collaborative* forms of governance (Blanco, 2015) on smart(er) cities, e.g., in terms of creating a higher urban quality of life. Ansell & Gash (2007, p. 2) define *collaborative governance* as a "governing arrangement where one or more public agencies directly engage non-state stakeholders in a collective decision-making process that is formal, consensus-oriented, and deliberative and that aims to make or implement public policy or manage public programs or assets". Examples include community policing, watershed councils, regulatory negotiation, collaborative planning, community health partnerships, and natural resource co-management.

In terms of governing, the emphasis is on creating a mutual cooperation system comprising local governments, research institutes, private companies, and citizens as well as different cities (C2C). The goal is to enable seamless data sharing and to find novel ways to address urban problems, e.g., by creating a cross-border testbed to verify SC services and infrastructure. One example is the European 'Urban Sharing Platform' (USP) funded by the EU's Horizon 2020 Research and Innovation Programme which led to innovation exchanges between the cities of London, Lisbon, and Milan: "The design of the platform is shaped by the commitment of Sharing Cities to open data, interoperability and integration (open APIs)" (http://www.sharingcities.eu/sharingcities/news/Simple-words-What-is-an-Urban-Sharing-Platform-Interview-with-Antony-Page-and-Jason-Warwick-Urban-DNA-WSWE-AWCH89). By sharing their own SC solutions with each other, participating cities become more efficient in aligning urban needs with SC technologies and services which in turn can lead to a better usage of city resources.

How effective is collaborative governance in relation to resolving complex SC problems such as climate change induced inner city floods, i.e., in situations where issues cannot be easily resolved or where the consultation process is (too) time consuming? How best to respond when participating entities such as individual activists, state agencies and private sector organizations can't find a consensus due to 'stakeholder fatigue'? Can 'innovative' governance combinations such as e-governance enhance transparency and accountability? How to manage powerful stakeholder groups that may seek to

---

Living Labs

A unique city-to-city (C2C) case example in terms of collaborative partnerships is Amsterdam's CITXL (The City Innovation Exchange Lab) which "creates social impact globally by inclusive experimentation, testing with the public in Living Labs and sharing helps cities identify common problems, co-develop solutions, identify technology and social impact to find the sweet spot for quick wins that make a difference in people's lives". Its facilitators conduct talks, walks and workshops (e.g., on 'responsible crowd sensing') and help cities to crowdsource solutions in communities (http://www.citixl.com/workshops/).

---

manipulate the overall governance process, mistrust, power imbalances, and cultural barriers?

Resource issues also affect governance agendas and outcomes. A case in point is the transformation of Medellín, a city of more than 2 million in Colombia with a troubled history of narcotics-related crime, poverty and despair (Freedman, 2019). Medellín's change makers put a premium on societal inclusiveness (e.g., qua neighbourhood meetings), and SC-related needs-based changes such as the construction of a gondola line to link the poor mountain communities to their jobs in the city induced by the communities themselves. *Did the city's resource limitations provide that extra incentive to ensure overall project success? A comparison that examines the differences and similarities between developed and underdeveloped cities would help to reveal the importance of SC governance context and specific governance modes vis-à-vis social, process and resource conditions as well as actual SC project outcomes.*

Strong SC governance has to ensure that there is public trust (Chan, 2019) in digital data sharing approaches used to achieve SC goals and that robust data protection rules are applied to SC technologies on the basis of codified frameworks (IMDA/PDPC, 2019).

4.3. *Behavioural Science Theories*

Since Thaler & Sunstein's (2008) publication of 'Nudge: Improving Decisions About Health, Wealth, and Happiness', there has been great interest across public policy domains in what drives the behaviour and decision-making logic of individuals or groups. Instead of traditional compliance methods such as education, legislation or enforcement, nudging puts emphasis on reinforcement and indirect suggestions as influence strategies. According to Thaler & Sunstein (2008, p. 6), a nudge is defined as "any aspect of the choice architecture that alters people's behaviour in a predictable way without forbidding any options or significantly changing their economic incentives." Examples of nudges are the deterrent disease pictures on cigarette packs encouraging smokers to reduce cigarette consumption through emotional responses or the rumble strip on highways that let drivers know if they are drifting out of the lane. In the context of SC, nudging is arguably a very effective approach that city leaders can use to support sustainability and liveability goals so that commuters modify transport-related choices aimed at minimizing congestion on the basis of personalised 'active' push notifications that nudge them to optimise individual personal routes without getting stuck in traffic (BVA Nudge Unit, 2019). *Evidence-based research on using behavioural design and SC apps to improve quality of urban life is still nascent.*

SC applications such as well-intended, analytical nudging efforts and associated issues such as concerns about unacceptable paternalism raise several managerial challenges for SC stakeholders be it city managers, SC businesses or citizens. One challenge is to skilfully manage the iterative behavioural design processes and systems testing for optimal behaviour-change purposes. While specialised behavioural design consulting firms such as the US BVA Nudge Unit or the UK Behavioural Insights Team (in short: 'Nudge Unit') have the required competencies to inform policy, improve public services, and deliver positive results for city dwellers (Quinn, 2018), it is rather unlikely that the ordinary civil servant in a SC unit commands such competencies. This also pertains the ethical and effective deployment of SC-related data-driven nudges and behavioural insights. To what extent government officials are supportive of nudge training programmes has to be examined. Another challenge is the need to effectively comply with data protection laws and consent frameworks. As Ranchordas (2019) has argued, employing

IoT, big data, and algorithms to nudge citizens into SC-befitting behaviour raises several legal and ethical issues such as data portability rights and the need to be transparent and beneficial to the public whenever nudging is used.

The trust implications of collecting and processing large amounts of personal and impersonal data to influence citizens' behaviour in smart cities are poorly researched. Digital trust refers to the confidence SC citizens have in an organization's ability to protect and secure data and the privacy of individuals. How do SC practitioners such as researchers, technologists, policymakers, corporations, and government entities ensure that SC technology applications are deployed in the interests of social good, rather than cause more public distrust? Another interesting question for further research is how effective data-driven SC nudges are in changing citizens' behaviour towards more sustainable cities and resource protection, e.g., in the area of offsetting one's personal carbon footprint.

4.4. Theories of Networks and Ecosystems

Networks are central to urban policy making and governance. They mobilise all sorts of resources and can enable a more plural, inclusive and participative approach to urban policymaking (Blanco, 2015). The ASEAN Smart City Network mentioned earlier is an example of a regional network aimed at connecting different smart cities ecosystem partners for knowledge sharing and to create business opportunities (https://www.smartcitiesnetwork.net/why-smart-cities-network). At supra-national levels, the 'C40 Cities' network connects almost 100 of the world's megacities to address climate change (https://www.c40.org/about).

Within cities, urban space is composed of many different networks: economic, social, political, technical and infrastructural (Pflieger & Rozenblat, 2010). *How these networks intersect at a given point with what kind of SC-related outcomes represents a poorly researched topic.* To ensure (technical) connectivity within the city, a smart connectivity infrastructure is needed with a blend of *network technologies* such as 4G LTE, 5G, and Wi-Fi, depending on the respective use cases. The IoT provides the infrastructure for communicating with sensors and other remote devices. *More research is necessary to examine how political decision-makers, political parties and political coalitions mobilise resources for different SC policy arenas.* As Blanco (2015) has argued, transformational smart city visions and success are influenced by the outcomes of political competition between alternative coalitions within a city. In that sense, 'collaborative governance' qua networking can have very positive effects on the realisation of SC initiatives. *To what extent this holds true for Asia's emerging cities of the future (e.g., under the conditions of resource limitations) in contrast to award-winning SCs in Europe or the US needs to be further examined.*

Relevant in terms of our discussion on network theories and collaborative stakeholder relationships in urban ecosystems are public-private partnership (PPP) projects. Take the case of the $1.33 billion Singapore Sports Hub, one of the world's first fully integrated sports, entertainment & lifestyle destination opened in 2015 (Lange *et al.*, 2018). Singapore's development progress has led to a high level of urban liveability, and the Government is determined to create a city where Singaporeans can 'live', 'work', 'learn' and 'play'. The development of this costly sports centre was made possible through a unique Public-PrivatePartnership (PPP) model but competing public and private sector priorities, and stakeholder alignment issues have plagued the island state's sporting crown jewel for some time. The facility is run by SportsHub Pte. Ltd., a consortium comprising four equity partners: InfraRed Capital Partners, Dragages Singapore, Cushman & Wakefield Facilities & Engineering, and Global Spectrum Asia. It has a 25-year contract with Sport Singapore (a statutory board under the Ministry of Culture, Community and Youth of the Singapore Government) to design, build, finance and operate the complex. In line with the public-private partnership agreement, the Singapore Government makes annual payments of $193.7 million to SHPL over a period of 25 years (from 2010 onwards) to run the Sports Hub. Tensions are caused by different preferences of governmental and business actors as well as 'unmet standards' (*Straits Times*, 2020). While some critics doubt that the PPP model is the right model for the Singapore Sports Hub, decisive performance monitoring (a typical weakness in other countries) by

Sport Singapore is one pragmatic approach to ensure that the Sports Hub evolves as originally envisaged.

Research on the performance effectiveness of SC-related PPPs is often hampered by the lack of financial data of P3s as a result of commercial confidentiality provisions. More research is needed on alternative PPP forms such as the public–private–community partnership (PPCP), in which both governmental and private players collaborate to improve cities, reducing or perhaps eliminating return of capital (ROC) and profit concerns.

Similarly, there is a need to assess the influence of the UN's 'United for Smart Sustainable Cities' (U4SSC) initiative within the Asian region. U4SSC is coordinated by the ITU along with several UN bodies, and provides SC leaders with guidance and advice along their smartness and sustainability pathways (https://www.itu.int/en/ITU-T/ssc/Pages/KPIs-on-SC. aspx). The U4SSC website features several valuable SC case studies, city snapshots, factsheets and verification reports (Dubai, Singapore, Valencia, Pully, Wels, etc. — with emphasis on KPI standards: http://www.itu.int/pub/T-TUT-SMARTCITY*).*

U4SSC

The "United for Smart Sustainable Cities" (U4SSC) is a UN initiative coordinated by ITU, UNECE and UN-Habitat, and supported by CBD, ECLAC, FAO, UNDP, UNECA, UNESCO, UNEP, UNEP-FI, UNFCCC, UNIDO, UNOP, UNU-EGOV, UN-Women and WMO to achieve Sustainable Development Goal 11: "Make cities and human settlements inclusive, safe, resilient and sustainable". U4SSC serves as the global platform to advocate for public policy and to encourage the use of ICTs to facilitate and ease the transition to smart sustainable cities (https://www.itu.int/en/ITU-T/ ssc/united/Pages/default.aspx).

Another promising area for further research is ecosystem theory (Moore, 1997; Williamsen & De Meyer, 2020). The term 'ecosystems' refers to "the complex of living organisms, their physical environment, and all their interrelationships in a particular unit of space" (https://www.britannica.com/science/ecosystem). Loosely coupled networks do matter when it comes to understanding the dynamics of smart cities, e.g., smart city initiatives aimed at making urban living safer and more sustainable by excelling in the six SC components introduced earlier: governance, the economy, mobility, environment, living and people.

A case in point is Berlin's 5.5ha European Energy Forum (Euref) campus (a business, research and education hub built on a former industrial site) that hosts several clean-energy-related companies and organisations such as the Green Garage, a cleantech accelerator that helps startups turn the climate challenge into a business opportunity. Another Euref tenant is InfraLab Berlin, a long-term co-working project of leading infrastructure and energy companies such as waste management firm Berliner Stadtreinigung (BSR), Berliner Verkehrsbetriebe BVG; Berlin's main public transport company) and Vattenfall (a major power company), to develop innovative smart-city solutions. One project under discussion is aimed at upgrading BVG's public bus fleet with moving sensors that scan the environment for necessary maintenance works in order to avoid costly spillovers of manholes after heavy rainfall. It is a pilot measure of Greenbox Global Holding GmbH (https://www.greenbox. global/) aimed at creating innovative value in the areas of environmental protection, infrastructure, energy supply and digitalisation. With Berlin as a reference case, the question arises how the InfraLab approach could be exported to (smart) cities in Asia.

The example underlines the benefits city leaders can gain when they utilise the expertise of various stakeholders across relevant industries, such as utilities, waste and recycling, telecommunications, high tech and so forth. As Claps (2017) has stressed, the power of the ecosystem 'will determine the ability of smart cities to realise the benefits of digital innovation'.

More research is necessary to answer the following research questions: What are some of the key factors that municipal leaders need to consider when selecting 'good' ecosystem partners? What are some of the good / best technology practices of aggregating and making sense of data collected from relevant organisations across the city ecosystem that can help users to do their jobs better? How best to turn analytical insights into relevant application processes in support of the six SC components, safely, ethically and efficiently?

Digital Innovation

To reduce the risk of flooding, the City of Buenos Aires in Argentina collected data from sensors in thousands of storm drains, maintained by the city public works department, to measure the direction, speed and level of water. These data were combined with weather forecast data from the Meteorological Service. City managers used the aggregated information to predict flood-prone areas in order to alert affected citizens. They also sent maintenance crews to affected neighbourhoods to clean storm drains. Through this digital innovation programme, the city managed to reduce flood damage to people and property from big storms (Claps, 2017; https://www.smartcity.press/climate-change-in-buenos-aires/).

4.5. *Theories of Technology Adoption and Business Model Innovation*

As indicated earlier, the adoption (Rogers, 2003) of SC technologies requires deep knowledge about the different SC technologies such as sensor technology or data protection, including associated use cases. To what extent city councils interested in implementing such technologies can appreciate the value and benefits of emerging SC technologies is a topic for further research. The job description of a SC specialist (see Appendix 1) indicates that the successful adoption cannot be taken for granted because the competency requirements are indeed very complex. Therefore, the question arises who and what drives the adoption of SC technologies in different contexts, e.g., developing vs. developed nation. *To what extent is the successful or unsuccessful implementation of SC components and applications contingent upon skilful technology push, organisational capabilities, leadership and managerial competencies, available resources, maturity level of beneficiaries/ users and so forth* (Godin, 2012; Teece, 2010; van den Ende & Dolfsma, 2005; Johnsen 1982; Wernerfeld, 1984)?

The success and failure of SC initiatives suggest that proactive and reactive mindset concepts (Bateman & Crant, 1993; Chen *et al.*, 2012), transformational leadership theory as well as business model

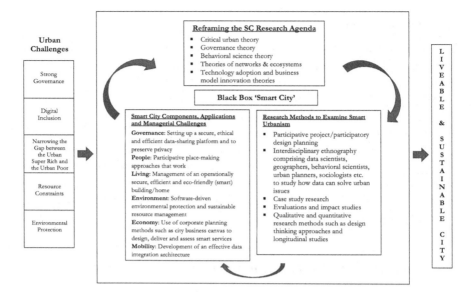

Figure 1.   A Framework for Future Smart City Research

innovation frameworks (Girotra & Netyessine, 2014; Osterwalder *et al.*, 2010; Bock & George, 2018; Taran *et al.*, 2019) have an important role to play in future SC research. A key question in this context is to what extent city councils, SC leaders and managers as well as other stakeholders do understand the business model innovation potential of *datafied* SC approaches. Many SC technologies are still in their pre-commercial stage, a great opportunity for extending existing theoretical frameworks about business models on these enablers qua empirical research. The same goes for research on the potential of blockchain and crypto to finance necessary investments in municipal SC infrastructures.

5. Conclusion

We set out to conduct a review of managerial challenges in relation to SC applications and to develop a framework for future research on Smart Cities in this paper. If there is a conclusion that can be drawn from this effort, it is the insight that (i) there is a multitude of more or less robust frameworks put forward by SC practitioners and (ii) that a grand research-driven SC theory has yet to be developed.

Due to the multiple components of the smart city model, 'strong' theorizing about SC is riddled with problems such as a seemingly general lack of theory in some of the SC frameworks, the 'technocratic' nature of the 'instrumented', data-driven city discourse or the imputed 'theory hostility' of some SC practitioners. Strategic SC frameworks by big technology firms often do not sufficiently consider the social, procedural or resource concerns raised by some of the SC components and applications. Emerging SC theories with great(er) explanatory potential are often based on narrow case study approaches and require further empirical tests in order to demonstrate their value.

Besides considerable ignorance about the actual mechanisms that make a city liveable, inclusive and sustainable, there are considerable managerial challenges (see Table 5) that impede the successful adoption of smart city applications as outlined above across all six SC components. In view of these challenges, we argue based on the following postulates that conceptual SC frameworks and further research can inform and enhance the job of SC leaders and managers tasked with the management of 'Smart Cities'.

5.1. *Postulates about the Benefits of Stronger SC Theorizing*

- Acknowledging that some of the key arguments of *critical urban theory* can support SC managers in designing and implementing more needs-based and socially inclusive SC approaches;
- A deeper engagement with *governance theory* can help SC managers to implement better approaches towards 'collaborative' or 'ICT-enabled', more transparent governance;
- Studying the main arguments of *nudge theory* can help SC managers to influence citizens to support sustainability and liveability goals;
- Knowing some of the facets of *network and ecosystem theories* can assist SC managers to better influence the set-up and expected

outcomes of multi-actor stakeholder ecosystems aimed at achieving SC goals (e.g., related to sustainable living or digitalisation);

- Acknowledging the premises of *technology adoption theory* and the power of connected systems can help SC managers to avoid that IoT projects fail at the PoC stage, e.g., by developing a dedicated, automated PoC platform;
- Examining how other cities have successfully used concepts of *business model innovation* can support SC managers to enhance the public planning and decision-making process behind smart(er) services.

The SC field is riddled with methodological issues such as lack of causality-oriented, empirical research with focus on the relationship between SC applications carried out by SC managers and leaders and the responses of citizens who are supposed to benefit in terms of impact. As argued by Lee *et al.* (2020), there is a need to deploy more appropriate research methods to examine smart urbanism such as interdisciplinary ethnography comprising data scientists, psychologists, behavioural scientists, sociologists, historians etc. to study how data driven SC applications can solve urban issues.

'Smart Cities' can contribute to well-being, inclusion, sustainability and resilience. The SC toolbox is well equipped. The challenge is to use the 'right' tool for the job and to know what is 'right' and what is unsound. Theory can help to avoid unsound action. This review has revealed that there are multiple theoretical avenues for future research pertaining 'critical urban theory', 'governance frameworks', 'behavioural science theory', 'theories of networks and ecosystems' or 'technology adoption and business model innovation theories', to further enhance our knowledge about effective SC approaches. In addition, other conceptual frameworks are worth to be explored to expand our knowledge about SC phenomena such as surveillance capitalism theory that questions the rationale of ICT, sensors, cyber-physical systems and analytics as 'good' SC applications; transformation theories related to city leaders and their complex SC tasks; marketing theories about successful SC city branding; strategic group formation to understand why some city councils succeed to become smart(er) while others fail; and many more.

Considerations of building more liveable and sustainable cities have the potential to improve the quality of urban life in line with the sustainable development goals/ESG. There is an urgent need to put more emphasis on the wellbeing of the urban poor in the world's mega cities who are (arguably) largely excluded from the SC discourse. To tackle the manifold challenges and opportunities that the Smart City discourse raises such as environmental threats, resource scarcity, social inequality, equitable use of SC technology and 'good' governance requires the inputs of experts from various fields and discipline areas. We hope that this article will be useful in stimulating more value-added, multi-disciplinary SC research to create sustainable, liveable urban communities where people are co-creators of SC policies.

References

Apte, M.P. (2017). Smart cities — A delusion of misplaced priorities. http://www.citymayors.com/development/smart-cities-india.html.

Australian Public Service Commission (APSC). Delivering Performance and Accountability. https://www.apsc.gov.au/delivering-performance-and-accountability.

Arbes, R. & Bethea, C. (2014). Songdo, South Korea: City of the future? *The Atlantic*, September 27. http://www.theatlantic.com/international/archive/2014/09/songdo-south-korea-the-cityof-the-future/380849/.

Ansell, C. & Gash, A. (2007). Collaborative governance in theory and practice. *Journal of Public Administration Research and Theory*, 18, 543–571.

Bateman, T.S. & J.M. Crant (1993). The proactive component of organizational behavior: A measure and correlates. *Journal of Organizational Behavior*, 14(2), 103–118. doi: 10.1002/job.4030140202.

Batty, M. (2012). Smart cities, Big data. *Environment and Planning B: Planning and Design*, 39, 191–193.

Bernt, M. (2012). The 'double movements' of neighbourhood change: Gentrification and public policy in Harlem and Prenzlauer Berg. *Urban Studies* 49(14), 3045–3062.

Blanco, I. (2015). Between democratic network governance and neoliberalism: A regime-theoretical analysis of collaboration in Barcelona. *City*, 44, 123–130.

Bock, A. & George, G. (2018). *The Business Model Book: Design, Build and Adapt Business Ideas That Drive Business Growth*. London: Pearson Business.

Brenner, N. & Theodore, N. (2002). *Spaces of Neoliberalism: Urban Restructuring in North America and Western Europe*. Boston, MA: Blackwell.

Brenner, N. (2004). *New State Spaces: Urban Governance and the Rescaling of Statehood*. New York, NY: Oxford University Press.

Brenner, N. (2009). What is critical urban theory? *City*, 13(2–3), 198–207.

Bristish City Council (n.d.). Corporate strategy 2017–2022. https://www.bristol.gov.uk/documents/20182/1188753/Corporate+Strategy+2017-2022+D5/c545c93f-e8c4-4122-86b8-6f0e054bb12d (accessed on 10 December 2020).

Business Dictionary. Description of smart city. www.businessdictionary.com/definition/smart-city.html (accessed on 2 December 2020).

BVA Nudge Unit (2019). Behavioral smart cities: A behavioral approach to improving urban efficiency and quality of life. https://bvanudgeunit.com/a-behavioral-approach-to-smart-cities/ (accessed on 7 December 2020).

Chan, D. (2019). Why and how public trust matters? Research Collection School of Social Sciences. https://ink.library.smu.edu.sg/soss_research/2915.

Chan, H.C. & Neo, H. (2018). 5 big challenges facing big cities of the future. https://www.weforum.org/agenda/2018/10/the-5-biggest-challenges-cities-will-face-in-the-future/.

Chen, Y., *et al.* (2012). Origins of green innovations: The differences between proactive and reactive green innovations. *Management Decision* 50(3), 368–398. doi: 10.1108/00251741211216197.

Christophers, B. (2011). Revisiting the urbanization of capital. *Annals of the Association of American Geographers*, 101(6), 1347–1364.

Chun, A. (2012). Ethnoscapes. In G. Ritzer (Ed.), *The Wiley-Blackwell Encyclopedia of Globalization* (9th edn.). John Wiley & Sons, Ltd.

Claps, M. (2017). Smart cities: The power of the ecosystem. https://www.smartcitiesworld.net/opinions/opinions/smart-cities-the-power-of-the-ecosystem (accessed on 10 December 2020).

Cohen, B. (2012). 6 key components for smart cities. www.ubmfuturecities.com/author.asp?section_id=219&doc_id=524053 (accessed on 2 May 2016).

Cohen, B. (2014). The 10 smartest cities in Europe. www.fastcoexist.com/3024721/the-10-smartest-cities-in-europe (accessed on 2 March 2016).

Cohen, B. & Muñoz, P. (2016). *The Emergence of the Urban Entrepreneur: How the Growth of Cities and the Sharing Economy Are Driving a New Breed of Innovators.* Santa Barbara: Praeger.

Cobb, S. (2019). Siegeware: When criminals take over your smart building. https://www.welivesecurity.com/2019/02/20/siegeware-when-criminals-take-over-your-smart-building/ (accessed on 9 December 2020).

Digital News Asia (2016). Singapore one of the most sustainable and smartest cities in the world. Report by *Digital News Asia,* 9 November.

Domaradzka, A. (2018). Urban social movements and the right to the city: An introduction to the special issue on urban mobilization. *VOLUNTAS: International Journal of Voluntary and Nonprofit Organizations,* 29, 607–620.

Eizenberg, E. (2012). Actually existing commons: Three moments of space of community gardens in New York City. *Antipode* 44(3), 764–782.

European Commission (2013). A digital agenda for Europe: A Europe 2020 initiative (Smart Cities). http://ec.europa.eu/digital-agenda/en/smart-cities (accessed on 10 July 2017).

EUR-SCM (2006). European Smart City Model. www.smart-cities.eu/model.html (accessed on 2 May 2016).

Ferro Guimarães, J.C., *et al.* (2020, April 20), Governance and quality of life in smart cities: Towards sustainable development goals. *Journal of Cleaner Production,* 253, Article 119926.

Fraser, N. (2014). Technology is not serving the ends it could serve in [rebuilding] public space. CCCB (Centre de Cultura Contemporania de Barcelona) Video Recording. https://www.cccb.org/en/multimedia/videos/nancy-fraser-technology-is-not-serving-the-ends-it-could-serve-in-rebuilding-public-space/229347.

Freedman, D.H. (2019). How Medellín, Colombia, became the smartest city. *Newsweek,* 18 November. https://www.newsweek.com/2019/11/22/medellin-colombia-worlds-smartest-city-1471521.html.

Girotra, K. & Netessine, S. (2014, July–August). Four paths to business model innovation. *Harvard Business Review.* https://hbr.org/2014/07/four-paths-to-business-model-innovation (accessed on 29 September 2020).

Godin, B. (2012). Innovation studies: The invention of a specialty. *Minerva,* 50, 397–421.

Goonawardene, N., Lee, P., Tan, H.-X., Valera, A.C. & Tan, H.-P. (2018). Technologies for ageing-in-place: The Singapore context. In T. Menkhoff, *et al.* (Eds.), *Living in Smart Cities: Innovation and Sustainability.* Singapore: World Scientific Publishing.

Gray, S., *et al.* (2017, April). Combining participatory modelling and citizen science to support volunteer conservation action. *Biological Conservation,* 208, 76–86.

Greenfield, A. (2013). *Against the Smart City.* New York: Do Projects.

Harris, A. & Sartipi, M. (2019). Data integration platform for smart and connected cities. In *SCOPE '19: Proceedings of the Fourth Workshop on International Science of Smart City Operations and Platforms Engineering,* April 2019, pp. 30–34. https://doi.org/10.1145/3313237.3313301.

Harvey, D. (1989). The urban experience. In *The Urban Process Under Capitalism: A Framework for Analysis* (Chapter 2, pp. 59–89). Baltimore: The Johns Hopkins University Press.

Harvey, D. (2013). *Rebel Cities: From the Right to the City to the Urban Revolution.* London: Verso.

Harvey, D. (2016). 'CCCB' (Centre de Cultura Contemporania de Barcelona) video recording. https://www.cccb.org/en/multimedia/videos/david-harvey-i-like-the-idea-of-an-urban-common-which-is-a-political-concept-which-says-that-this-space-is-open-for-all-kind-of-people/229344.

Hoang, T.-A., Cohen, W., Lim, E.P., Pierce, D. & Redlawsk, D. (2013). Politics, sharing and emotion in microblogs. In *Proceedings of the 2013 IEEE/ACM International Conference on Advances in Social Networks Analysis and Mining (ASONAM 2013),* Niagara Falls, 25–28 August 2013 (pp. 282–289). New York: ACM. http://dx.doi.org/10.1145/2492517.2492554.

Hoe Sue Fern (2018, May). From liveable to lovable city: The role of the arts in placemaking Singapore. *Social Space,* 23, 9–21. https://socialspacemag.org/from-liveable-to-lovable-city-the-role-of-the-arts-in-placemaking-singapore.

IBM (2012, March). How to transform a city: Lessons from the IBM smarter cities challenge. *IBM Smarter Cities White Paper.* http://asmarterplanet.com/files/2012/11/Smarter-Cities-WhitePaper_031412b.pdf.

IBM-SC. Smarter cities: The insight to identify, transform and progress. www.ibm.com/smarterplanet/sg/en/smarter_cities/overview/ (accessed on 2 May 2016).

IMDA/PDPC (2019). Trusted data sharing framework. https://www.imda.gov.sg/-/media/Imda/Files/Programme/AI-Data-Innovation/

Trusted-Data-Sharing-Framework.pdf (accessed on 15 December 2020).

ITU Academy (n.d.). Key performance indicators (KPIs) and standards for smart sustainable cities smart sustainable cities training programme. Module SSC-3. https://www.itu.int/en/ITU-D/Regional-Presence/AsiaPacific/Documents/Module%203%20 Smart%20Sustainable%20Cities%20KPIs%20Draft%20H.pdf (accessed on 31 December 2020).

Kitchin, R. (2014). The real-time city? Big data and smart urbanism. *GeoJournal*, 79, 1–14.

Khoo, T.C. (2016). Building a liveable city: The Singapore experience. In F.L. Lye & J. Wong (Eds.), *The Challenge of Making Cities Liveable in East Asia.* Singapore: World Scientific Publishing.

Letsch, C. (2013, May 31), Turkey protests spread after violence in Istanbul over park demolition. *The Guardian.* London. (accessed on 8 December 2013).

Lange, K., Chan, C.W. & Rathbone, M. (2018). Singapore's sports hub — Achieving 'smart living' status through public-private partnerships (PPP). In T. Menkhoff, S.N. Kan, H.-D. Evers & Y.W. Chay (Eds.), *Living in Smart Cities: Innovation and Sustainability.* Singapore: World Scientific Publishing.

Lee, J.Y., Woods, O. & Kong, L. (2020). Towards more inclusive smart cities: Reconciling the divergent logics of data and discourse at the margin's. *Geography Compass,* 14(9), 1–12.

Legatum. Legatum prosperity index. www.li.com/programmes/prosperity-index (accessed on 2 May 2016).

Libelium (2020). 50 sensor applications for a smarter world. https://www.libelium.com/libeliumworld/top-50-iot-sensor-applications-ranking/.

Luo, X. & Yang, J. (2019). A survey on pollution monitoring using sensor networks in environment protection. *Journal of Sensors,* Article ID 6271206. https://doi.org/10.1155/2019/6271206.

McGuirk, P.M. (2004). State, strategy, and scale in the competitive city: A neo-gramscian analysis of the governance of 'global sydney. *Environment and Planning A,* 36(6), 1019–1043.

Monetary Authority of Singapore (MAS) (2020, December 7), Speech by Mr Heng Swee Keat', Deputy Prime Minister, Coordinating Minister for Economic Policies and Minister for Finance, for Singapore FinTech Festival x Singapore Week of Innovation & TeCHnology (SFF X SWITCH) 2020. https://www.mas.gov.sg/news/speeches/2020/speech-by-mr-heng-swee-keat-at-sff-x-switch-2020 (accessed on 11 December 2020).

Mayor of London (2018). Mayor launches roadmap to make London the world's smartest city. https://www.london.gov.uk/press-releases/ mayoral/mayor-launches-smart-london-plan (accessed on 14 December 2020).

Menkhoff, T., Kan, S.N., Evers, H.-D. & Chay, Y.W. (2018). *Living in Smart Cities: Innovation and Sustainability.* Singapore: World Scientific Publishing.

Moore, J.E. (1997). *The Death of Competition: Leadership and Strategy in The Age of Business Ecosystems.* New York: HarperBusiness.

Ng, Y.-D. (2020, December 1), How smart cities can serve citizens. *PhysOrg.* https://phys.org/news/2020-12-smart-cities-citizens.html.

Ng, C.K. (2019). What does it take to be a smart nation? *GOVINSIDER,* 28 August 2018. https://govinsider.asia/connected-gov/ng-chee-khern-what-does-it-take-to-be-a-smart-nation-sndgg/ (accessed on 11 December 2020).

Novotný, R., Kuchta, R. & Kadlec, J. (2014). Smart city concept, applications and services. *Journal of Telecommunications System & Management,* 3(2), 1–8 https://www.hilarispublisher.com/open-access/smart-city-con-cept-applications-and-services-2167-0919-117.pdf.

Osterwalder, A., Pigneur, Y. & Clark, T. (2010). *Business Model Generation: A Handbook For Visionaries, Game Changers, and Challengers*, Strategyzer Series, Hoboken, NJ: John Wiley & Sons.

Pak, J. (2018). Creative placemaking as a policy and a practice of urban regeneration in Singapore: Negotiating power relations and forging partnerships in civic society. Asia Research Institute Working Paper Series No. 268. National University of Singapore. https://ari.nus.edu. sg/wp-content/uploads/2018/10/wps18_268.pdf.

Pflieger, G. & Rozenblat, C. (2010). Introduction. Urban networks and network theory: The city as the connector of multiple networks. *Urban Studies,* November 16. https://journals.sagepub.com/ doi/10.1177/0042098010377368 (accessed on 10 December 2020).

Phang, S.-Y. (2018). Alleviating urban traffic congestion in smart cities. In T. Menkhoff, *et al.* (eds.), *Living in Smart Cities: Innovation and Sustainability.* Singapore: World Scientific Publishing.

Poole, S. (2014). The truth about smart cities: In the end, they will destroy democracy. *The Guardian,* 17 December.

Quinn, B. (2018). The 'nudge unit': The experts that became a prime UK export. *The Guardian,* 10 November. https://www.theguardian.com/

politics/2018/nov/10/nudge-unit-pushed-way-private-sector-behavioural-insights-team (accessed on 9 December 2018).

Rahman, K. (2018). A case study of the DTSS — Changi water reclamation plant project. In T. Menkhoff, *et al.* (Eds.), *Living in Smart Cities: Innovation and Sustainability.* Singapore: World Scientific Publishing.

Rathorea, M.M., *et al.* (2018, June). Real-time secure communication for smart city in high-speed big data environment. *Future Generation Computer Systems*, 83, 638–652.

Ranchordas, S. (2019). Nudging citizens through technology in smart cities. *International Review of Law, Computers & Technology*, 34(3), 254–276. https://www.tandfonline.com/doi/full/10.1080/13600869.2019.1590 928?af=R.

Rogers, E. (2003). *Diffusion of Innovations* (5th edn.). New York: Simon and Schuster.

Rutland, T. (2013). Activists in the making: Urban movements, political processes and the creation of political subjects. *International Journal of Urban and Regional Research*, 37(3), 989–1011.

Schmidt, A.T. & B. Engelen (2020). The ethics of nudging: An overview. *Philosophy Compass*, 15(4). https://onlinelibrary.wiley.com/doi/full/ 10.1111/phc3.12658 (accessed on 9 December 2020).

Sennett, R. (2012). No one likes a city that's too smart. *The Guardian*, 4 December.

Shelton, T., Zook, M. & Wiig, A. (2015). The actually existing smart city. *Cambridge Journal of Regions, Economy and Society*, 8, 13–25.

Shepard, W. (2017). China's most infamous 'ghost city' is rising from the desert. *Forbes*, 30 June. https://www.forbes.com/sites/ wadeshepard/2017/06/30/ordos-chinas-most-infamous-ex-ghost-city-continues-rising/?sh=5ffbe3e76877#6aff89a46877/2017/06/30/.

Smart Nation Singapore (n. d.). Initiatives. https://www.smartnation.gov. sg/what-is-smart-nation/initiatives (accessed on 4 January 2021).

Straits Times (2020, March 7), Singapore sports hub fined over unmet standards. https://www.straitstimes.com/politics/singapore-sports-hub-fined-over-unmet-standards (accessed on 10 December 2020).

Tan, K.G., *et al.* (2012). *Ranking the Liveability of the World's Major Cities. The Global Liveable Cities Index (GLCI).* Singapore: World Scientific Publishing.

Taran, Y., Goduscheit, R.C. & Boer, H. (2019). Business model innovation — A gamble or a manageable process? *Journal of Business Models*, 7(5), 90–107.

Teece, D.J. (2010). Technological innovation and the theory of the firm: The role of enterprise-level knowledge, complementarities, and (dynamic) capabilities. In *Handbook of the Economics of Innovation*, Vol. 1 (Chapter 16). Elsevier B.V.

Temperton, J. (2015, August 17). One Nation under CCTV: The future of automated surveillance (Wired UK). www.wired.co.uk/news/archive/2015-08/17/one-nation-under-cctv (accessed on 2 March 2016).

Thaler, R. & Sunstein, C. (2008). *Nudge: Improving Decisions about Health, Wealth, and Happiness*. New Haven, CT: Yale University Press.

Timeus, K., Vinaixa, J. & Pardo-Bosch, F. (2020). Creating business models for smart cities: A practical framework', *Public Management Review*, 22(5), 726–745: Special issue: Management, Governance and Accountability for Smart Cities and Communities. Guest editors: Giuseppe Grossi, Albert Meijer and Massimo Sargiacomo. https://www.tandfonline.com/doi/full/10.1080/14719037.2020.1718187.

Townsend, A.M. (2013). *Smart Cities: Big Data, Civic Hackers, and the Quest for a New Utopia*. New York: W. W. Norton & Company.

UN (2014). World urbanization prospects (Highlights). https://esa.un.org/unpd/wup/ (accessed on 18 July 2017).

URBACT — European Exchange and Learning Programme Promoting Sustainable Urban Development (2019). How participatory placemaking can help URBACT local groups to develop urban actions for public spaces in our cities. https://urbact.eu/how-participatory-placemaking-can-help-urbact-local-groups-develop-urban-actions-public-spaces-our (accessed on 9 December 2020).

van den Ende, J. & Dolfsma, W. (2005). Technology-push, demand-pull and the shaping of technological paradigms — Patterns in the development of computing technology. *Journal of Evolutionary Economics*, 15, 83–99. doi: doi.org/10.1007/s00191-004-0220-1.

Wamsler, C., Alkan-Olsson, J., Björn, H., Falck, H., Hanson, H., Oskarsson, T., Simonsson, E. & Zelmerlow, F. (2020). Beyond participation: When citizen engagement leads to undesirable outcomes for nature-based solutions and climate change adaptation. *Climatic Change*, 158, 235–254.

Williamson, P. & De Meyer, A. (2020). *Ecosystem Edge: Sustaining Competitiveness in the Face of Disruption*. Stanford, CA: Stanford Business Books.

Appendix 1: Job Description of a Smart City Specialist

Job Description

The International City/County Management Association (ICMA) is seeking a Smart Cities Specialist to support the USAID/Guatemala Creating Economic Opportunities (CEO) project.

Background:

The USAID/Guatemala Creating Economic Opportunities (CEO) project supports economic growth, private sector development, competitiveness, and job creation in Guatemala. It strengthens the promotion of investment and trade, catalyzes productive infrastructure; develops the workforce of Guatemala, and improves the business enabling environment. A central objective of the CEO Project is to strengthen the private sector as a growth engine to reduce poverty, improve living conditions, and create sustainable economic opportunities in Guatemala for Guatemalans. **By focusing on the country's secondary cities as natural investment and growth platforms, cultivating partnerships between stakeholders in the public, private and civil society sectors, as well as emphasizing an ecosystem conducive to innovation and entrepreneurship, the CEO project is playing a key role in job creation, facilitating investment, and allowing prosperity beyond the interior of the country.** As part of its focus on innovation, CEO has identified a municipality whose political leadership has committed to transforming the municipality into a Smart City which can generate safer space with better local services and an environment of innovation that encourages creative solutions, promotes job creation and reduces inequalities.

Job Description:

The Smart Cities Specialist will facilitate the transformation of a small Western Highlands municipality in Guatemala into a Smart City. The Specialist will travel to Guatemala to gather information, interview stakeholders and discuss options with local government officials and staff. S/he will develop an Action Plan with key strategies and the model that will drive the transition of the municipality into a Smart City. The strategy will include concrete actions to attract investment, promote job creation and improve infrastructure to enhance the quality of life for citizens and encourage business growth.

(*Continued*)

(*Continued*)

What you'll do:

Collect and analyze information and initial **plans** developed by the municipality related to their vision/conceptualization of Smart City concepts.

Analyze the **municipality's governance and planning capacities** (Municipal Development Plans, Urban Plans, Operational Plans, among others).

Analyze the status and potential of municipal services that could be improved using **Smart City technologies/concepts**, identifying the critical challenges and opportunities for moving forward with the implementation of a Smart City Action Plan.

Identify synergies and **strategic alliances with the public and private sectors** (local, national and foreign) that could be harnessed in support of the **transformation** of the municipality into a Smart City.

Identify viable **technological solutions** for the territory to respond to the problems or opportunities identified.

Develop an **action plan** for transforming the municipality into a Smart City to improve the efficiency of municipal management and promote local economic development in a systematic and sustainable way.

Identify short, medium and long-term actions (and investment plans) for the municipality to facilitate its **transformation**, along with clear implementation methodologies, indicators and timelines.

Identify a **portfolio of potential investment projects** and potential resources for **financing**.

What you need to be successful in this role:

Education

Master's Degree in **information technology, urban/city planning, economics, government, business administration, systems engineering, sustainable development, international trade, or related careers.**

(*Continued*)

(Continued)

Knowledge, Skills and Abilities
Individuals must have experience in the development and implementation of Smart City technologies/projects and the minimum qualifications indicated below:

- Five or more years of recent experience (preferably public sector) providing guidance to cities to lead to their transformation into a Smart City.
- Experience in the **conceptualization and execution of economic development projects with Smart City concepts**.
- Experience **implementing urban design projects** with the use of **technological tools**.
- Experience in **developing strategies for Smart Cities with multidisciplinary groups**.
- Experience working in developing/transitional countries to implement Smart City concepts.
- Fluency in spoken and written Spanish.

Source: https://www.devex.com/jobs/smart-cities-specialist-658562.

Chapter 13

Nurturing Youth Climate Action: A Blue Carbon Perspective

Thomas Menkhoff and Kevin Cheong

It was announced at the COP27 Climate Conference in Sharm El-Sheikh (Egypt) in November 2022 that Amazon and Conservation International (with support from the Singapore Economic Development Board) will establish an International Blue Carbon Institute in Singapore. As Minister for Sustainability and the Environment, Grace Fu rightly pointed out during the launch in the Singapore Pavilion that "South-east Asia, with its vast stretches of mangroves and coastal ecosystems, has tremendous potential for blue carbon initiatives that will also support environmental protection, biodiversity conservation, and livelihoods and heritage of local communities" (The Straits Times, New blue carbon institute to be launched in Singapore, 15/11/2022).

As educators involved in sustainability and smart city-related courses, we certainly welcome this new initiative. However, while the concept of blue carbon has gained recognition among scientists, policymakers, and environmentalists, it is not yet widely understood by the general public.

In fact, from an educational standpoint, blue carbon is both an opportunity (e.g., in terms of mangrove regeneration) and a pedagogical challenge. Just like carbon trading or ESG reporting, 'blue carbon,' a term coined in 2009 to highlight the contribution of coastal vegetated ecosystems to climate change mitigation, is still a closed book for many youths. However, given the alarming new climate change projections, there should be a greater sense of urgency in raising awareness and building capacity so that our youths can appreciate their essential role in fighting climate change.

As educators, we are compelled to ask: Can we do more?

Blue carbon refers to the carbon dioxide stored in coastal and marine ecosystems like mangrove forests, seagrass meadows, and tidal marshes. These ecosystems sequester and store large quantities of carbon in both plants and sediments, thus playing an essential role in climate adaptation and mitigation. Mangrove trees store three to five times more organic carbon than tropical upland forests, according to climate tech start-up Handprint Tech. Besides protecting communities from heavy flooding associated with rise of sea levels, mangroves prevent soil erosion and promote spawning of sea life.

While it is easy to present and explain such facts to students in a classroom setting, the more challenging task is to really engage learners so that they acknowledge the worsening climate change problem and take concrete action.

According to research conducted by the Singapore-ETH Centre in collaboration with the Centre for Liveable Cities, participating in information and activation events, such as pop-up booths and tree-planting events, can increase social resilience against climate change-related risks. The term 'social resilience' refers to the ability of social entities to adapt to challenges, such as the recent pandemic or climate change-related risks, and to cope with adverse events. In our view, a rather under-explored question is the extent to which local university students experience climate anxiety and recognise the need to build personal climate resilience in preparation for the future.

In an attempt to nurture youth climate action awareness, emphasising blue carbon education, we recently collaborated with Singapore's Gardens by the Bay in a multidisciplinary, project-based SMU-X course 'Innovations for Asia's Smart Cities' that exposed undergraduates to the Kingfisher Wetlands, a 1.5-ha man-made waterbody and garden that opened in 2021. Here, the Gardens' team has planted more than 200 native mangroves in the Kingfisher Wetlands, including the Nipah Palm and the endangered Upriver Orange Mangrove, and is conducting research to examine the potential of such mangrove habitats as a long-term nature-based solution for capturing and storing carbon in other urban areas.

One promising result of studies conducted by the Gardens (which is collaborating with the National University of Singapore's Centre for Nature-based Climate Solutions and environmental consultant DHI Water and Environment) is that the carbon content of the pond's underwater soil sediment is comparable with what is found in other natural wetland habitats, underscoring their importance as good carbon sinks.

Our students took part in a hands-on blue carbon-related mangrove monitoring activity within the Kingfisher Wetlands pond led by the Gardens staff. Through this 'get-your-feet-wet activity,' the youths learnt to appraise the health of our tropical, salt-tolerant wetland mangrove trees (also called halophytes) which prevent shoreline erosion by absorbing storm surge impacts in case of extreme weather events. Mangrove monitoring is critical in diagnosing causes of environmental stress and in preventing mangrove degradation. The healthier the mangrove roots, the better their ability to stabilise shorelines and function as carbon sinks.

To support the Gardens' sustainability strategy, one student group also set out to raise awareness — among their fellow Singapore youths — of the urgency of nature-based solutions to climate change, issues aimed at protecting, sustainably managing, and restoring coastal ecosystems of mangroves. Activating their mental gears, they came up with (i) an engagement campaign and digital outreach materials for a self-guided learning tour around the Kingfisher Wetlands pond; (ii) gamified information boards that would quiz visitors about 'blue carbon' through crossword puzzles; and (iii) stations where visitors can learn more about carbon molecules and pledge commitment towards performing simple eco-friendly acts. We are hopeful that visitors to the Gardens will soon experience their ideas in action.

Judging from their course feedback, the hands-on mangrove monitoring activity in the muddy waters of the Kingfisher Wetlands pond was a hit with our city youths. While the novelty factor may have played a part, as educators, we are optimistic that the rare immersive exposure to our mangrove — a precious but disappearing jewel — has engaged our youths at a deeper level and has planted a seed of motivation.

Perhaps more research is needed to help us ascertain the extent of impact of such blue carbon-focused environmental excursions to our coastal shoreline (or to islands such as Pulau Ubin, where NParks and OCBC Bank are building a mangrove park to increase Singapore's capacity for carbon storage) have on transforming our youth into 'valuable contributors to climate action' to coin a UN term.

Or even better, Singapore's institutions of higher learning with their different discipline areas, specialisations, and eco laboratory equipment could join forces and co-create novel climate education approaches so that more young Singaporeans become agents of change, green entrepreneurs, and blue innovators to accelerate climate action.

Printed in the United States
by Baker & Taylor Publisher Services